THE USES OF ANTIQUITY

AUSTRALASIAN STUDIES IN HISTORY AND PHILOSOPHY OF SCIENCE

VOLUME 10

General Editor:

R. W. HOME, *University of Melbourne*

Editorial Advisory Board:

W. R. ALBURY, *University of New South Wales*
D. W. CHAMBERS, *Deakin University*
R. JOHNSTON, *University of Wollongong*
H. E. LE GRAND, *University of Melbourne*
A. MUSGRAVE, *University of Otago*
G. C. NERLICH, *University of Adelaide*
D. R. OLDROYD, *University of New South Wales*
E. RICHARDS, *University of Wollongong*
J. J. C. SMART, *Australian National University*
R. YEO, *Griffith University*

The titles published in this series are listed at the end of this volume.

THE USES OF ANTIQUITY
The Scientific Revolution and the Classical Tradition

Edited by

STEPHEN GAUKROGER

*Department of Traditional and Modern Philosophy,
University of Sydney, Australia*

KLUWER ACADEMIC PUBLISHERS
DORDRECHT / BOSTON / LONDON

Library of Congress Cataloging-in-Publication Data

```
The Uses of antiquity : the scientific revolution and the classical
  tradition / edited by Stephen Gaukroger.
       p.    cm. -- (Australasian studies in history and philosophy of
  science ; v. 10)
    Includes index.
    ISBN 0-7923-1130-2 (HB : acid-free paper)
    1. Science, Ancient.  2. Science--Philosophy--History--17th
  century.   I. Gaukroger, Stephen.  II. Series.
  Q124.95.U83  1991
  509--dc20                                                    90-26819
```

ISBN 0-7923-1130-2

Published by D. Reidel Publishing Company,
P.O. Box 17, 3300 AA Dordrecht, Holland.

Sold and distributed in the U.S.A. and Canada
by Kluwer Academic Publishers,
101 Philip Drive, Norwell, MA 02061, U.S.A.

In all other countries, sold and distributed
by Kluwer Academic Publishers Group,
P.O. Box 322, 3300 AH Dordrecht, Holland

Printed on acid-free paper

All Rights Reserved
© 1991 Kluwer Academic Publishers
No part of the material protected by this copyright notice may be reproduced or
utilized in any form or by any means, electronic or mechanical,
including photocopying, recording or by any information storage and
retrieval system, without written permission from the copyright owner.

Printed in The Netherlands

FOREWORD

The institutionalization of History and Philosophy of Science as a distinct field of scholarly endeavour began comparatively early — though not always under that name — in the Australasian region. An initial lecturing appointment was made at the University of Melbourne immediately after the Second World War, in 1946, and other appointments followed as the subject underwent an expansion during the 1950s and 1960s similar to that which took place in other parts of the world. Today there are major Departments at the University of Melbourne, the University of New South Wales and the University of Wollongong, and smaller groups active in many other parts of Australia and in New Zealand.

'Australasian Studies in History and Philosophy of Science' aims to provide a distinctive publication outlet for Australian and New Zealand scholars working in the general area of history, philosophy and social studies of science. Each volume comprises a group of essays on a connected theme, edited by an Australian or a New Zealander with special expertise in that particular area. Papers address general issues, however, rather than local ones; parochial topics are avoided. Furthermore, though in each volume a majority of the contributors is from Australia or New Zealand, contributions from elsewhere are by no means ruled out. Quite the reverse, in fact — they are actively encouraged wherever appropriate to the balance of the volume in question.

R. W. Home
General Editor
Australasian Studies in History and Philosophy of Science

TABLE OF CONTENTS

STEPHEN GAUKROGER / Introduction: The Idea of Antiquity	ix
KEITH HUTCHISON / Copernicus, Apollo, and Herakles	1
JOHN SUTTON / Religion and the Failures of Determinism	25
JAMIE C. KASSLER / The Paradox of Power: Hobbes and Stoic Naturalism	53
UDO THIEL / Cudworth and Seventeenth-Century Theories of Consciousness	79
ALEXANDER JACOB / The Neoplatonic Conception of Nature in More, Cudworth, and Berkeley	101
JAMES FRANKLIN / The Ancient Legal Sources of Seventeenth-Century Probability	123
KIRSTEN BIRKETT and DAVID OLDROYD / Robert Hooke, Physico-Mythology, Knowledge of the World of the Ancients and Knowledge of the Ancient World	145
JOHN GASCOIGNE / 'The Wisdom of the Egyptians' and the Secularisation of History in the Age of Newton	171
GARRY W. TROMPF / On Newtonian History	213
NOTES ON CONTRIBUTORS	251
INDEX OF MYTHICAL AND HISTORICAL FIGURES	253

STEPHEN GAUKROGER

INTRODUCTION: THE IDEA OF ANTIQUITY

Antiquity was conceived of, put to use, and reassessed in various ways in natural philosophy and what might broadly be termed metaphysics in the period between Copernicus and Newton. The papers in this collection deal with questions about the symbolic and polemical uses of antiquity, with antiquity as a fund of ideas or as a source of evidence, and above all with the ways in which an image of antiquity was constructed and put to use in contemporary debates. As an introduction to these questions, I want to set the scene by drawing attention to a feature of pre-Enlightenment thought about antiquity that, in its more extreme form, can only be termed a Christian-allegorical reading of antiquity. It is a reading in which ancient thought is construed as leading inexorably to, and finding its culmination in, Christian dogma. This makes the interpretation of antiquity a question charged with controversy, and no more so than in the period under consideration in this collection.

During the Enlightenment, the period from Copernicus to Newton was pictured in terms of a struggle between those who relied exclusively on the storehouse of ancient wisdom, and those who managed to break free from the shackles of the past and start out on a fresh path, the struggle being won decisively by the latter. Voltaire summed up the victory in his article on Job in his *Dictionnaire Philosophique*, in the claim that there wasn't a book in science in his day 'that is not more useful than all the books of antiquity'. The failings of this picture of what came to be termed 'the quarrel between the ancients and the moderns' are now well known. It fails to face up to the way in which the past was used in the sixteenth and seventeenth centuries. It fails to take account of the complex ways in which the past was conceived, and especially the way in which it was extensively drawn upon to defend or attack contemporary views both by the followers of the 'ancients', and the moderns alike. It also fails to take account of the fact that there was considerable uncertainty as to how the past was to be understood, on its significance, on what kinds of evidence could be drawn upon, as well as on the vexed question of cultural diffusion in antiquity. These latter questions, about how antiquity was to be conceived, had a very significant bearing on how considerations about the past were used in argument, and indeed the two sets of questions are in many respects intimately bound together.

As an illustration, take the case of Joseph Glanville. Glanville was a staunch modernist, and was — with Cowley, Evelyn, and Sprat — one of the principal apologists for the Royal Society in the Restoration. He was an ardent follower of Bacon, an empiricist, and (despite his belief in witchcraft) a proponent of scepticism in all matters not scientifically demonstrable. In *A Letter to a Friend Concerning Aristotle* (1665), he spells out his modernist opposition to Aristotle in a way which, far from pitting the moderns against antiquity, uses antiquity to rebut Aristotle:

> ... the reverence I have to the more *antient Sages*, which *Aristotle* frequently *traduced*, and unworthily *abused*, animated me to more Severity against *him*, than upon another occasion had perhaps been so *pardonable* and *becoming*. And that *Aristotle* dealt so *invidiously* with the Philosophers that were before him, will not need much proof to one, that is but indifferently acquainted with his writings. The great Lord *Bacon* hath particularly charged him with this unworthiness in his excellent *Advancement of Learning*, wherein he says, that '*Aristotle* as though he had been of the race of *Ottomans*, thought he could not reign, except that the first thing he did, he kill'd all his Brethren.' And elsewhere in the same Discourse 'I cannot a little marvel at the Philosopher *Aristotle*, that proceeded in such a spirit of difference and contradiction to all *Antiquity*, undertaking not only to frame new words of Science at pleasure, but to confound and extinguish all the *antient Wisdom*, insomuch that he never names any *Antient* Author, but to confute or reprove him' consonant whereunto are the observations of *Patricius* that he carpes at the *Antients* by name in more than 250 places, and without name in more than 1000. [H]e reprehends 46 *Philosophers* of worth, besides *Poets* and *Rhetoricians*, and most of all spent his spleen upon his excellent and venerable Master *Plato*, whom in above 60 places by name he hath contradicted. And as *Plato* opposed all the *Sophisters*, and but two *Philosophers*, viz. *Anaxagoras* and *Heraclitus*; so *Aristotle* that he might be opposite to him in, *this* also, oppos'd all the *Philosophers*, and but two *Sophisters* viz, *Protagoras* and *Gorgias*. Yea, and not only *assaulted* them with his arguments, but *persecuted* them by his *reproaches*, calling the *Philosophy* of *Empedocles*, and all the Antients *Stuttering; Xenocrates*, and *Melissus*, *Rusticks; Anaxagoras*, *simple* and *inconsiderate*; yea, and all of them in a heap, as *Patricius* testifies, *gross Ignorants*, *Fools* and *Madmen*. How fit then think you is it that the World should now be obliged to so *tender* and *awful* a respect to the *Libeller* of the most Venerable *Sages*, as that it should be a crime next *Heresie* to endeavour, though never so *modestly*, to weaken his *textuary* and *usurp'd authority*? when my *veneration* of the greater *Antiquity* extorted from me those strictures against the *proud Antagonist* of all the *ancient* and more *valuable Wisdom*?[1]

INTRODUCTION: THE IDEA OF ANTIQUITY

In short, Aristotle is a *parvenu*, aggrandizing himself at the expense of his more illustrious predecessors. What Glanville seeks to put in the place of Aristotelian natural philosophy is a form of atomism, but his vindication of it lies not in its subsequent success as a natural philosophy in his own day, but in its claim to have been the first or original philosophy, this priority somehow being linked to its present success:

> That the *Aristotelian* was not the antient *Philosophy*, but the *Corpuscularian* and *Atomical*, which to the great hinderance of Science lay long buryed in *neglect* and *oblivion*, but hath in these latter Ages been again restored to the *light* and it's deserv'd *repute* and *value*. And that the *Atomical Hypothesis* was the *First* and most *Antient*, of which there is any memory in *Physiology*, is notoriously known to all, that know the Age of *Democritus*; who was one of those Four *Sages* that brought the learning of the *Aegyptians* among the *Grecians: Orpheus* bringing in *Theology; Thales* the *Mathematicks*; our *Democritus*, natural *Philosophy*; and *Pythagoras* all *Three*, with the *Moral*. Now the learning of the *Aegyptians* came from the *Chaldeans*, and was convey'd to *them*, as some learned men affirm, by *Abraham*, who was of kin to *Zoroaster* to great *Chaldean Legislatour* and *Philosopher*; which *Zoroaster* lived 290 years after the *Flood*, and as *Pliny* saith, was the Schollar of *Azonaces*, whom *Antiquaries* affirm to have been of the Schoole of *Sem* and *Heber*. The *Atomical Philosophy* then coming from the *Aegyptians* to the *Grecians*, and from the *Chaldeans* to *them*; is without doubt of the most *venerable Antiquity*; and the *Aristotelian* a very *novelty* in compare with that *grey Hypothesis*: at the best, a *degeneracy* and *corruption* of the most *antient Wisdom*. Yea, and 'tis the complaint of several learned Men, which whoever knows any thing of *Aristoteles Sectators* will justifie, That the Modern *Peripateticks* have as farr receded from *his* sense, as from the *Truth* of *Things*. For it hath been the Fashion of his *Interpreters* both *Greeks, Latins*, and *Arabians*, to form whole Doctrines from *catches* and *scraps* of sentences, without attending to the *analogy* and *main scope* of his Writings. From which method of interpretation hath proceeded a *spurious medly* of *nice, spinose* and *useless notions*, that is but little of kin to *Aristotle* or *nature*. So that whatever of genuine *Aristotelian* is in those *works* that bare *his* name; There's little of *Aristotle* in his *Schools*. And 'tis no indignity to *Antiquity* or the *Stagyrite*, to oppose the *corruption* and *abuse* of *both*. And to endeavour to restore the *Antients* to their just *estimation*, which hath been usurp't from them by a *modern* and *spurious* Learning. And though I grudge not *Aristotles* esteem while it is not prejudicial to the respect we owe his *Betters*; yet I regret that excessive and undue *veneration* which fondly sets him so much above all the more valuable Antients.[2]

This kind of attempt to find the most ancient sources of modern doctrines is a comon one in the seventeenth century, and it has seemed to many commentators that Hermeticism provides the key to understanding what lies behind such a project.[3] But the reconstruction of ancient sources was not confined to Hermeticism at all, and the kind of project in Georg Horn's *Historiae Philosophicae* (Leiden, 1655) for example — where all philosophy is traced back to Adam, the various philosophical schools or sects being simply a result of the fall — is just a more orthodox version (at least in Protestant countries) of the kind of enterprise fostered by the Hermeticists. Moreover, Isaac Casaubon's demonstration, in 1614, that at least a very significant part of the Corpus Hermeticum dated from the Christian era, seriously undermined the Hermetic attempt to establish the identity of Platonism and Christianity. It is true that writers such as Cudworth questioned a number of aspects of Casaubon's dating: he refuses to accept, for example, that the presence of elements of Greek philosophy precluded an earlier dating, on the grounds that since 'Pythagorism, Platonism and the Greek Learning in general, was in great part derived from the Egyptians, it cannot be concluded, that whatever is Grecanical, therefore was not Egyptian.'[4] But Cudworth does not rely on Hermetic sources as evidence, as Ficino has done in his *Theologia Platonica* (1469–74), the pioneering Hermetic work of the Renaissance. This is instructive, for both the *Theologia* and Cudworth's *True Intellectual System of the Universe* were concerned to combat materialist and naturalist philosophies which defend such doctrines as the eternity of the world, the mortality of the soul, and the regulation of nature by blind necessity. Moreover, both see the sources of these doctrines in Epicurean atomism, or in Stoicism or variants of Aristotelianism, both rely on Plato and especially Plotinus, and both provide complex genealogies for their own views. But whereas Ficino can simply take it that the Corpus Hermeticum provides conclusive evidence that Platonism and Christianity were essentially the same system, expressed differently, Cudworth is forced to be much more circumspect about the historical evidence. Moreover, his core metaphysical/natural-philosophical thesis, that an atomist natural philosophy entails a commitment to dualism, is supported both on historical grounds, by an elaborate reconstruction of early thought on which it transpires that the Pythagoreans were the first atomists, and on quite independent conceptual grounds, by means of an argument designed to show that an atomist construal of matter as essentially inert requires us to postulate active agencies in nature over and above matter.

Given this last argument, one may wonder why Cudworth devotes such an immense amount of attention to the historical case: just as one may wonder why Glanville — albeit no doubt influenced by his friend Cudworth — chooses to defend atomism, the scientific success of which was secure at the time he was writing, in the way he does. The theological acceptability of atomism was certainly at issue, and Glanville and Cudworth are just as certainly concerned with this, but what has the tracing of the doctrine back to the pre-Christian era to do with its theological credentials?

To find the answer to this question, we need to go beyond Hermeticism to the more general current of thought about antiquity in which Hermeticism was but one alternative. In this respect, it is interesting to note that we find disputes as early as the Patristic period over which pagan sources are closest to Christianity: Lactantius, for example, glorifies Hermes as a Gentile prophet whereas Augustine maintains that his foreknowledge of Christianity was due to communion with devils.[5] A deeply-embedded assumption in such disputes is that Christianity somehow pervades the whole of the pre-Christian era, and the thought and beliefs of this era are subjected to a kind of allegorical reading in order to yield or reveal what are often marvellous anticipations of Christianity. Such an allegorical reading of the past is not peculiar to the Church Fathers: there is evidence to suggest that Homer was being read allegorically as early as Plato's time, and the third-century Neoplatonists Porphyry and Plotinus developed this into an art form, subjecting Homer's works to a detailed allegorical reading which enabled them to present him as a sage with revealed knowledge of the fate of souls and of the mystical structure of the universe.[6] They used this allegorical reading to attack Christianity, and it is not surprising that the Church Fathers should reply in kind: and with a vengeance, for the Patristic project was a far more ambitious and all-encompassing one than that of the Neoplatonists.

In the Patristic period, we witness a gradual 'Christianization' of philosophy (metaphysics, natural philosophy, ethics, etc.), begun by the early Fathers and brought to completion by Augustine. It amounts to a total translation of philosophy into Christian terms. Christianity is conceived of as the final form of philosophy. Using the language of the classical philosophers to formulate their theology, they attempted to show that Christianity was able to answer all the questions of classical metaphysics. In general terms, not only does Christianity supplement classical philosophy, it appropriates the teachings of this philosophy, denying that they were ever the property of the ancients in the first place, and it construes every philosophical question in terms of Christian teaching. This appropriation of earlier thought by Christianity made it possible for it to present itself as the final answer to what earlier philosophers were striving for, and in a number of cases it was strikingly successful in this respect: the great ease with which it transforms one of the central aims of Hellenistic philosophy (whether Stoic, Epicurean, Sceptic or Platonist), namely that of transcending the flux and disorder of life and the achievement of peace of mind (*ataraxia*), into Christian terms is nothing short of remarkable.

In some areas, however, this appropriation was acutely problematic, and this is especially so on the question of the vexed relation between Christianity and its religious forebear, pre-Christian Judaism. In the early centuries of Christianity, there was for a while a close contest between those who saw Christianity as the true development of the religious precepts contained in the Old Testament, and those who saw the two as completely opposed. The Manichaeans and Gnostics had held that the Law of the Old Testament had been abrogated by the coming of Christ, and that as a consequence the Old Testament itself should be discarded. The second-century Gnostic Marcion, in his *Antitheus,* had set out the moral and theological discrepancies between the Old Testament and the Gospels, and argued that the former was the record of the Jewish God of Hate, something which was now superseded by the message of the God of Love in the Gospels. Allegorical reading of the Old Testament was explicitly ruled out by Marcion as an attempt to save something which was in fact wholly alien to Christian tradition. The God of the Old Testament was still accepted as a God, however, with the result that two independent realms of evil and good were postulated, with independent Gods ruling over these. Augustine had in fact been a Manichaean in his early 20s,[7] but quickly became one of its fiercest critics: it was crucial to Christianity, as he later construed it, that there was only one God, and that this God was the God of the Old Testament as well as the New. In shifting from Manichaeism to his mature position, he was strongly influenced by Neoplatonism, with its conception of the 'One' as incorporeal, immutable, infinite, and the souce of all things: a conception which completely contradicts the Manichaean/Gnostic idea that there could be a God who was vengeful and spiteful. In his discussion of Plato in Books 8 to 10 of the *City of God*, he speculates whether Plato could have had some knowledge of the Hebrew scriptures, and he suggests that the God of the Neoplatonists is the same as that of Christianity, and even that they speak, albeit in a confused way, of the Trinity. Yet these same

Neoplatonists cannot reach God. They mistakenly believe that they can reach Him by purely intellectual means, whereas in fact He can only be reached through the sacraments, which were instituted with the Incarnation of Christ. For Augustine, the superiority of Christianity over ancient philosophies and over the contemporary rivals of Christianity lay in the institution of the sacraments. But it is not so much that Christianity is ancient philosophy plus the sacraments; a more accurate way of putting it would be to say that ancient philosophy is Christianity minus the sacraments. Christianity is the culmination of all previous philosophical reflection and religious belief, something that can be glimpsed by the appropriate allegorical readings of the ancient philosophers and sages just as much as it can by the allegorical reading of the Old Testament: and part of the error of the Manichaeans and Gnostics was precisely their refusal to join in such an allegorical reading, and instead to take the Old Testament literally.

The problem of the literal or allegorical reading of the Old Testament, I am suggesting, brings with it the more general problem of the interpretation of ancient thought and belief. The procedure of allegorical interpretation does not appear to have originated with Christian readings of the Old Testament, as I have indicated, and in any case Augustine's reading of Plato, for example, is just as allegorical as any of his readings of the Old Testament.

By the end of the Middle Ages, the theology of which was dominated by Augustine,[8] this problem resurfaced, and the question of the relation between Christianity and its religious and philosophical predecessors opened up again, albeit in a new way. During the Reformation, abuse within the Church stimulated a nostalgic desire for a return to earlier times when Jesus' simple message had been understood without the interpolations of Medieval Christianity. The 'return to Scripture' immediately opened up the question of the relation between the Old Testament and the Gospels, as the project was that of reconstructing Christianity on the basis of a reading of the New Testament that was free from the corruptions introduced by the interpretations of the Medieval Church, and the way to do this was to read the Gospels against the background of a more literal understanding of the Old Testament, an understanding often aided by reliance on the Rabbinic tradition, despite the rampant anti-Semitism of the time.[9] The project was of course a dangerous one, with radicals like Michael Servetus arguing that such corruptions began as early as the Council of Nicaea

(325) and the (in his view) clearly polytheistic doctrine of the Trinity. Others, more cautiously, saw the problems beginning later, perhaps as late as the ninth century. But irrespective of when the corruptions were thought to have begun, this line of thought places a new and heavy responsibility on the interpretation of antiquity. Antiquity now begins to be seen not just as a precursor of Christianity, but as *a source of clarification about the very nature of Christianity.*

'Antiquity' here covers not just pre-Christian Judaism, but those classical philosophies from which Christianity borrowed so much in formulating its basic doctrines. Because Christianity had now to be 'rediscovered', as it were, by studying its origins, antiquity had to be reassessed. And on this reassessment turned the correct understanding of Christianity. Given this, it is not surprising that, especially in Protestant countries, the question of the interpretation of antiquity is a key one, and one charged with controversy.

NOTES

[1] Joseph Glanville, *A Letter to a Friend Concerning Aristotle*, appended to *Scire/i tuum nihil est: or, The Author's Defence of the Vanity of Dogmatizing* . . . , London (1665), pp. 84—5. The reference to 'Patricius' is to Francesco da Cherso Patrizi (Franciscus Patritius) whose erudite *Discussiones peripateticae* (1581) is the source of the criticism that Aristotle took his philosophy from those he attacked.

[2] *Ibid.*, pp. 89—90. The reconstruction of ancient thought offered here derives from Cudworth, even though his main treatment of it, in his *True Intellectual System*, was not to appear for another thirteen years.

[3] The book that really established this interpretation was Frances Yates' immensely influential *Giordano Bruno and the Hermetic Tradition*, London (1964). Yates' enthusiastic defence of the role of Hermeticism has recently been subjected to a number of challenges; see, for example, the discussion in the editor's introduction to Brian Vickers (ed.), *Occult and Scientific Mentalities in the Renaissance*, Cambridge (1984).

[4] Ralph Cudworth, *The True Intellectual System of the Universe*, London (1678), p. 326.

[5] See Yates, *Giordano Bruno*, op. cit., Ch. 1.

[6] See Robert Lamberton, *Homer the Theologian*, Berkeley (1986).

[7] See Ch. 5 of Peter Brown, *Augustine of Hippo*, London (1967).

[8] See the exemplary discussion in Steven Ozment, *The Age of Reform*, New Haven (1980), Ch. 2.

[9] See Jerome Friedman, *The Most Ancient Testimony*, Athens, Ohio (1983) on the use of the Rabbinic tradition in the sixteenth century.

KEITH HUTCHISON

COPERNICUS, APOLLO, AND HERAKLES

Copernicus' personal seal (Fig. 1) is an image taken from classical antiquity. It depicts a near-naked man standing in a distinctive pose, with something flexible slung over his shoulder, while he holds up, and probably plays, a lyre. In an analysis of this seal, Mossakowski identifies the 'man' as Apollo, the principal solar deity of late antiquity, and argues that the image is an emblem of cosmic concord:[1] its use by the astronomer expresses his belief that the sun binds the universe together in some sort of harmonious unity. It expresses, in other words, the philosophical values already recognised as generating Copernican dissatisfaction with traditional Ptolemaic astronomy.

The argument which Mossakowski presents to support his interpretation is thoroughly satisfactory. He looks partly at contemporary Renaissance examples of similar emblems, and partly at ancient sources for the image. We know that Copernicus was educated within the

Fig. 1. Copernicus' personal seal, reprod. from Mossakowski, 'Symbolic Meaning,' *op. cit.* note 1, p. 452.

Renaissance humanist tradition and shared that tradition's great confidence in the wisdom of the classical world. So we can reasonably presume that the meaning attached by Copernicus to his seal was similar to that attached to comparable images in the ancient world, especially if it is evident that the ancient meanings are reflected in other Renaissance usages. Mossakowski cites, for example the well-known frontispiece to Franchinus Gaforus' *Practica Musice* of 1496, explicitly depicting a lute-playing Apollo in his classical role as leader of the nine Muses. These musical sisters had often functioned in ancient sources to represent the harmony of the universe, and Gaforus' picture overtly reproduces a classical belief that the principal components of the universe — the earth, the stellar sphere, and the seven planets — were correlated with the Muses. The planetary orbits are explicitly shown separated by musical intervals, as in classical accounts of the harmony of the spheres, and a banner announces that it is 'the power of Apollo's will [which] enlivens the whole circle of these Muses.' Copernicus' seal, we conclude, is a shorthand version of the complex of ideas set out more fully in the Gaforus frontispiece, a condensation of the whole diagram to the figure of Apollo at the top.[2]

The purpose of the present paper is to refine Mossakowski's analysis. I argue for an apparently different conclusion: that Copernicus' seal is an emblem of the mythical hero-become-god Herakles.[3] But Herakles himself, I also argue, had acquired distinct solar overtones by the time of the Renaissance, and I note further that he was a traditional classical emblem of cosmic harmony. So my alternative interpretation of the seal is not as different as it might initially seem. The seal more accurately represents, I further suggest, something which lies behind both Apollo and Herakles — the seemingly common hero of a whole family of ancient combat myths, where harmony and order is created and maintained by the defeat of a chaos-threatening monster.

This refinement of Mossakowski's Apollo interpretation is an important matter, and not simply an attempt to 'put the record straight.' For if we recognise that Copernicus' figure is as much Herakles (or perhaps also Theseus or Perseus or David or St. George, etc.) as it is Apollo, we are better placed to perceive that the astronomer has chosen an image for his seal which is very similar to emblems commonly used by the near-contemporaries of Copernicus and his followers. Variants of his emblem — especially those which use Herakles unambiguously — can be found, for example, amid the decorations on early modern centres of political authority.[4] Such repetitions of the image have the potential to add considerable depth to our understanding of the attractions of Copernican astronomy, and our noting them helps us to explore affinities between Copernican philosophical values, and the wider values of Renaissance society. For these heroic figures were certainly used to express social and political ideologies, and especially (it would seem) those which stressed the necessity for strong central government to maintain *social* harmony. So we can begin to see Copernicus' seal as a statement of social attitudes, social attitudes which may reasonably be suspected of generating the philosophical values which led to dissatisfaction with conventional cosmology. Useful support is thus given to a thesis I have argued elsewhere, that it was changing social attitudes in early modern Europe which made the Copernican account of the planetary motions seem more attractive than that inherited from Ptolemy. Harmony and coordination, especially that created by a single dominating influence, was deemed a more attractive feature of any system of interacting objects in the seventeenth century than it had been in the fifteenth.[5] This aesthetic judgment was apparently expressed by emblems like that chosen by Copernicus.

An easy entrée to this re-interpretation of Copernicus' seal is provided by the structure of Mossakowski's own argument. He identifies the figure in the seal as Apollo, both directly, by noting similarities in the iconography of other Apollo representations, and indirectly, by revealing the evident interpretive success that such an assumption will lead to. We can, however, do the same thing with Herakles, and the direct identification can be done somewhat better.

For the direct evidence specifically noted by Mossakowski is an ancient Apollo described in Furtwängler's classic survey of ancient gems.[6] This precise citation is supplemented by a general reference to some other studies of similar artforms, by Lippold and Richter. Perusal of these sources certainly confirms Mossakowski's claim: the image is undeniably classical, and it was used repeatedly to refer to Apollo.[7] But these same sources do not indicate that the image was used *exclusively* for Apollo, and show indeed that it was used for Herakles as well. Against the two precedents provided by Furtwängler for identifying the seal as Apollo, can be cited *four* for an identification with Herakles, and a similar ambiguity occurs with Lippold and Richter.[8] Indeed, in the totality of Mossakowski's ancient sources, there are only two Apollo representations which are particularly close to the precise iconography of the Copernican seal, and exactly the same number of similarly close Herculean figures can be found: see my Fig. 2 and Fig. 3. To these can be added a source not used by Mossakowski, some important coins of the first century BC which depict Herakles in a pose similar to that of Copernicus' musician: see Fig. 4.[9] So the identification with Apollo is clearly unwarranted on the basis of ancient iconography alone — *a marginally stronger case can be made for Herakles.*

When we turn to ancient literary evidence — not overtly used by Mossakowski since Apollo's connection with the lyre is so well-known — a similar conclusion can also be reached: Herakles is again portrayed as a musician or as an associate of the Muses. Much indirect evidence of this (together with a partial explanation) will be presented later in this study, but some very explicit cases can be mentioned now. The final lines of Ovid's *Fasti*, for instance, describe Herakles as 'twang[ing] his lyre',[10] and refer to a well-known Roman temple, containing statues of the Muses and a musical Herakles, erected by Fulvius Nobilior

Fig. 2. Ancient Herakles, from Furtwängler, *Antiken Gemmen, op. cit.* note 6, I, pl. XXXIV—25.

Fig. 3. Ancient Herakles, from Furtwängler, *Antiken Gemmen, op. cit.* note 6, I, pl. LVII—10.

Fig. 4. Roman Coin (minted ca. 67 BC by Pomponius Musa) depicting *Hercules Musarum*, and Apollo, from Roscher, *Lexikon, op. cit.* note 11, vol. I, col. 2972.

around 189 BC. Ovid again mentions this temple in his *Art of Love*; Suetonius notes that it was restored during the reign of Augustus; Macrobius cites it briefly in the Saturnalia; and Plutarch is presumed to be discussing it when (in his *Roman Questions*) he asks the rhetorical question: 'Why did Herakles and the Muses have an altar in common?' It is this temple which is referred to in the coins noted in the last paragraph.[11]

Music furthermore is explicitly associated with the education of the trainee-hero. This is an important component of the ancient mythology, as Herakles' education was eventually quite widely portrayed as a model for the inculcation of exemplary virtue, and it was Herakles' moral strengths which were especially emphasised in Renaissance repetitions of the mythology.[12] Both Plato and Aristotle recommended that music be part of the training of young aristocrats, and such training was particularly associated with stringed instruments. For in classical times (and in the Renaissance) wind instruments are commonly seen as undignified (if not subversive) and unsuitable for use by the nobility.[13] But Herakles ends up fighting with his teacher Linos, and clubs him to death with the lyre. As also happens with Apollo, mythographers thus present the hero's canonical weapon (the club) as like a musical instrument — both of them then, are equally devices for creating harmony. Apollodorus' account of the murder of Linos, furthermore, is immediately followed by the story of Herakles' defeat of the lion of Cithaeron, whose skin becomes the hero's canonical costume: this lion's name, however, is almost the same as the standard Greek word for the lyre [*kithara*, cf. 'guitar']. Linos is elsewhere associated with Apollo, either as son (in one case with mother Urania, the Muse with special care of astronomy), or as enemy. Like another musician, the flautist Marsyas, he is killed by Apollo for challenging the latter's musical supremacy.[14]

Similarly, the monster Cacus who fights Herakles at the lecture site of Rome while the hero is returning from Spain with Geryon's cows, is believed to have developed from a mythical Etruscan seer who was represented as looking like Apollo (also a prophet), and who similarly played the lyre. He too had close connections with Marsyas. So Herakles' feat of strength again has harmonic overtones, and resembles a musical contest like those engaged in by Apollo. Virgil's account of the conflict, as told to Aeneas by King Evander, ends with the company celebrating Herakles' cycle of labours in song. 'All the woodland rang in harmony [*consonat*] with the gay sound and the hills echoed it back.'[15]

In some versions of the murder of Linos, it is Eumolpus who teaches Herakles music. Linos teaches him his letter, for Linos is presented not only as 'the first ['among the Greeks'] to discover the different rhythms and song,' but also as 'the first to transfer [the Phoenician letters] into the Greek language, [and] to give a name to each character, and to fix its shape.' The implication of this story is that Herakles is particularly skilled with words, and he would thus be associated with the Muses in their capacity as patronesses of poetry, literature and philosophy, as well as music. So when Plutarch answers the question we heard him asking a few paragraphs ago (p. 5), it is Herakles' literary skills that are cited as the reason for sharing a temple with the Muses: Herakles is said to have taught the alphabet to early inhabitants of the site of Rome.[16]

Hermes is, however, the standard patron of philosophy and language, just as Apollo is the conventional leader of the muses, so Herakles evidently becomes a partial duplicate of Hermes, as he also does of Apollo. Hermes is sometimes credited with the invention of astronomy, and we will see later (p. 14) that this honour is also often given to Herakles.[17] A discussion in Lucian, especially important because of its repeated citation by Renaissance authors, illustrates this tendency to substitute Herakles for Hermes. He tells us that some 'connect eloquence not with Hermes ... but with the mightier Heracles ... who is Eloquence personified.' They 'refer the achievements of ... Heracles, from first to last, to his wisdom and persuasive eloquence. His shafts ... are no other than his words: swift, keen-pointed, true-aimed to do deadly execution in the soul.' For this reason, they depict Herakles as a wrinkled old man, dragging 'after him a vast crowd of men, all of whom are fastened by the ears with thin chains' held by Herakles' tongue.[18] Hermes furthermore is an important musician: it is in fact he who invented Apollo's lyre, as well as Pan's pipes, and his Caduceus is a standard emblem of concord.

So ancient sources clearly leave a Herculean interpretation of Copernicus' musician open to us. This interpretation can be greatly strengthened by seeking Renaissance parallels to the imagery used by Copernicus. For Mossakowski cites no precise analogues here, and relies on the superficially different image of Apollo as leader of the Muses, as in the Gaforus frontispiece discussed on p. 2 above. But again, we find that Apollo is not alone in this role — early modern sources repeat the ancient identification of Herakles as a duplicate leader of the Muses.

'One of the three crucial Renaissance mythographies', Giraldi's 1548 *De Deis Gentium*, for example, notes Herakles' Greek epithet *Musagetes* (= 'leader of the Muses'), and cites Eumenius' discussion of the temple in Rome in explanation. Vossius' *De Theologia Gentili*, the 'most important work of comparative mythology in the 17th century' mentions the connection briefly, and the annotated version of Alciati's *Emblemata*, 'the prototype for a profoundly characteristic Renaissance genre', mentions it three times.[19]

Cartari's *Imagini de gli dei*, 'the most popular of the three major [Renaissance] mythographies', includes its discussion of Herakles within the section on Mercury [= Hermes], noting that the two are little different, and introducing the hero's eloquence (and *prudenza*) as the first of his attributes to be discussed. A woodcut of the image described by Lucian is included in the illustrated editions. This would appear to be a good example of a phenomenon mentioned briefly above, the great emphasis placed on Herakles' moral and intellectual qualities in Renaissance Europe, where many political leaders (like the d'Este Ercoli) fostered an identification of the advantages of their rule with the virtues of the ancient hero. Cartari only mentions the Muses briefly, but an annotation (by Pignoria) in the 1615 and 1647 editions expands on this theme:[20]

> Not far removed [from the eloquent Herakles and Mercury] is Herakles Musagetes, the so-called leader [*guida*] of the Muses ... [A]s Eumenius the Rhetor writes, the peace [*quieta*] of the Muses, together with the quality [*valore*] of their voice and their song, needs defending by Herakles ... Eumenius based his ideas on the example of Fulvius who put up the temple in the circus [at Rome] to Herakles and the Muses ... So Herakles is sometimes seen with a lyre [*Cithara*] and a plectrum in his hand, as in a most beautiful cameo of the former Patriarch of Aquileia, and in medallions [coins?] of the Pomponia family.

Pignoria glosses this passage with a picture (my Fig. 5) of Herakles-the-musician taken from one of these ancient medallions, in the precise pose of the Copernican seal. This was a particularly appropriate image for representing Herakles' intellectual powers, for stringed instruments had apparently become a standard emblem of rhetoric in Pignoria's time.[21]

Such explicitly musical depictions of Herakles must have been reasonably familiar in early modern Europe, for other examples can be

 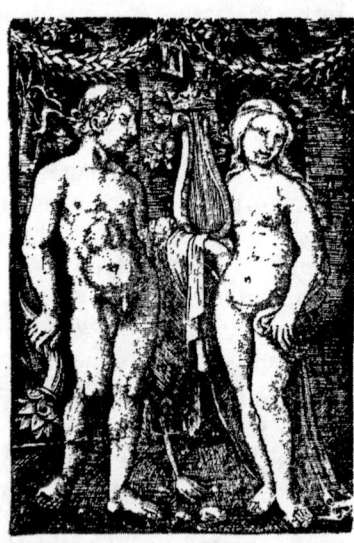

Fig. 5. Herakles playing the lyre, from the 1647 edition of Cartari, *Imagini*, *op. cit.* note 20, p. 341.

Fig. 6. Altdorfer (1480—1531), 'Hercules and a Muse,' *op. cit.* note 22.

found. Copernicus' near-contemporary, the German engraver Albrecht Altdorfer, did a picture of 'Hercules and a Muse', (my Fig. 6) which, although different in form from the seal we are attempting to identify, still portrays Herakles as holding a lyre, and is much closer to the Copernican image than the Renaissance models cited by Mossakowski. The Master of the Die similarly portrays Herakles, in the mid-sixteenth-century, as a protector of the Muses, who maintains concord among the nine sisters, by dispelling Envy from their Temple. The iconography here is quite unlike the Copernican seal, but a particularly close parallel can be found in an engraved 'Man Playing the Lyre', by the Master's Italian contemporary, Enea Vico (my Fig. 7); and Beger, considerably later includes a couple of equally close examples in his *Hercules Ethnicorum*.[22]

So there is good direct evidence — iconographic and literary, ancient and modern — for identifying Copernicus' picture as an emblem of Herakles, and the iconographic evidence is considerably stronger than the competing evidence cited by Mossakowski. What about the indirect evidence? If we assume the figure is Herakles, can we make good sense of Copernicus' use of it? Again, we find Herakles to be as plausible a choice as Apollo — for Herakles imitates Apollo in becoming a solar demigod, and has furthermore a special attachment to astronomy, being often identified as the source of human understanding of the stars. And as noted below, Herakles can also be identified as an emblem of cosmic

Fig. 7. Vico, 'Man Playing a Lyre', *op. cit.* note 22, mid-16c.

concord: the combats which form the core of the traditional Herculean stories were often portrayed in ancient sources as struggles on behalf of the harmony of the universe. A Herculean interpretation then, is just as satisfactory as an Apolline one.

A nice illustration of an ancient tendency to identify Herakles with the sun-god (rather than simply with the more standard planet Mars) is provided by the ancient coins used to confirm the Roman worship of *Hercules Musarum*. As indicated in my Fig. 4, some of these have Apollo depicted on the reverse side, while Du Choul's 'enormously influential' mid-fifteenth century study of the imagery of the classical gods reproduces a comparable Greek medal (my Fig. 8), with Herakles on one side and the fall of Phaeton, depicted against the background of the zodiac on the other.[23] Macrobius too is explicit about the identification. Herakles, he tells us, in a passage much noted by Renaissance mythographers,

> does not differ in essence from the sun, for he is that power of the sun which gives to the human race a valour like that of the gods ... Hercules is also believed to have slain the Giants in defense of heaven ... and it was the sun that exacted the due penalty from them by the destructive power of its heat. That Hercules is indeed the sun is clear from his very name, for the derivation of his Greek name "Heracles" is obviously ... "Pride of the Air" — and what, pray, is the pride of the air but the light of the sun ...?

Some Egyptian 'religious observances' point to the same conclusion, he continues, and so do the mythical details of the battle between King

Fig. 8. Greek medallion depicting Herakles and the sun chariot, from du Choul, *Religion des Romains* (1556), *op. cit.*, note 23, p. 180.

Theron of Spain and Herakles. Nonnos takes a similar view, describing 'starclad Herakles' as 'leader of the stars' and 'prince of the universe', and identifying him with both Helios and Delphic Apollo, as well as with various other gods.[24]

The fact that Nonnos wishes to identify Herakles with other gods beyond Apollo/Helios detracts somewhat from the value of his evidence. For it argues against the view that there is some special link between Herakles and solar Apollo. A similar view could be taken of the Macrobius discussion just quoted, for this is taken from a text which is arguing in favour of solar monotheism: Macrobius identifies many other gods with the sun as well.

Yet as the coins noted above indicate, ancient mythology does seem to have recognised a particular relationship between Herakles and the sun, and this special relationship has been acknowledged by a few modern commentators, who support it in part by citing the particularly telling belief that Herakles' twelve labours represented the annual motion of the sun through the signs of the zodiac. Examination of the sources cited for this belief, however, reveals that little ancient basis for it has been produced,[25] though the idea was certainly alive in the Renaissance as we see below. A closer discussion here is clearly desirable.

Plutarch does mention in passing an Egyptian belief that Herakles had his dwelling in the sun, and there is further evidence of Herakles' solar connections in the surviving accounts of the exploits of Herakles' important Mesopotamian prototype, Gilgamesh, where linkages with the sun are frequently made.[26] When noting the similarity with Herakles, Graves (in the most extended modern discussion of this issue familiar to me) describes Gilgamesh as 'connected with the progress of the sun around the zodiac', and repeatedly says much the same of Herakles, describing him at one stage as 'Lord of the Zodiac'. He cites an unspecified passage in Servius as evidence for this idea, but also notes a tradition of celebrating Herakles' birthday once every month, a number of specific correlations between individual labours and zodiacal constellations, plus a story that Herakles lost three hundred and sixty of his allies from Cleonae in a war linked by mythographers with the solar constellation Leo.[27] So while his claim certainly has some plausibility, it still remains ambiguously supported.

Renaissance texts mention Macrobius as an ancient source for the idea, but (as suggested in note 25) I doubt any passage can be found in his extant writings more explicit than that already quoted. Further evidence can however be found elsewhere, in a small number of ancient passages which describe the activities of the sun in characteristically Herculean terminology as 'labours'. One of these passages, Ovid's account of the myth of Phaeton, explains this terminology to us: Phoebus attempts to persuade his over-confident son to withdraw his request to take control of the sun's motion for a day, by pointing out the various difficulties of the job. There is a steep climb in the morning, until dizzy heights are reached at midday, followed by a precipitous descent: the wilful horses are difficult to control, and the ever-present motion of the celestial sphere needs to be fought against. Furthermore, 'you must make your way,' he tells his son (drawing a clear parallel between the yearly passage round the zodiac, and adventures similar to those of Herakles), through dangerous ambushes and among monstrous wild beasts: even though you keep to the path and do not wander [*nulloque errore*, 1.79] from it, still you will have to go past the horns of the hostile Bull, past the Thracian archer, and the jaws of the raging Lion, past the Scorpion's cruel pincers, whose sweeping embrace threatens you from one quarter, while the clutching claws of the Crab attack you from another.

'Enough [!],' he soliloquises later, after the death of his son, in obvious reference to the above remarks; 'from time's beginning has my lot been unrestful; I am weary of my endless and unrequited [labours] [*sine honore laborum*, 1.387]'. This same terminology is applied to the sun, in Virgil's description of the feast welcoming the Trojans to Carthage, towards the end of book I of the *Aeneid*. A local bard Iopas entertains the company with a cosmological song, and Virgil tells us that he 'makes the hall ring with his golden lyre. He sings of the wandering moon [*errantem lunam*], and the [labours] of the sun [*solisque labores*]'.[28]

Modern commentators, however, do not recognise this line as a reference to the belief that the motion of the sun through the sky is like Herakles' cycle of labours, and follow instead a long tradition of treating the *solis labores* as eclipses: Lewis and Short even cite this line as evidence that one of the meanings of the latin noun *labor* is 'eclipse', but evidence for this is relatively thin, and contrary evidence is quite strong.[29] In fact, Virgil's near-contemporaries Manilius, Lucan and Cicero both use the labour metaphor in astronomical contexts where it is clear an eclipse cannot be referred to. Lucan clearly describes the motion of the sun as a labour; Cicero contrasts the wandering of the planets with the laborious daily motion of the stellar sphere; and Manilius' *Astronomica* uses *labores*, almost as a technical term, to refer to the astrological 'activities' of the constellations of the zodiac, a usage which Lewis and Short only mention under the Greek loan-word *athlon*. Manilius himself tell us that his term is a translation of *athla* ('labours', the plural of *athlon*), but these words are apparently not found in any extant Greek astronomical works. They are however clearly related to *athlos* (cf. 'athletics'), one of the standard Greek words for Herakles' *labours*: it was Herakles indeed who was often seen as having started the Olympic games, in commemoration of his final victory over Augeas.[30]

A number of pre-twentieth-century commentators have certainly read Virgil's line as referring simply to the solar motions. The sixteenth century English translation by Phaer and Twyne, in particular, renders the phrase as 'the daily toile' 'of the Sonne', a translation which is quite inconsistent with the eclipse interpretation, but makes perfect sense when read in the light of the Ovid passage.[31]

The issue cannot be decided here, but it must be recognised that there are other reasons for presuming a reference to Herakles — even if the 'labours' in question really are eclipses. All three of the Cicero, Ovid, and Virgil passages just discussed exploit the contrast between 'wandering' and 'labouring', and this is a contrast which is frequently found elsewhere in ancient stories with obvious Herculean resonances. Gilgamesh's journeys are repeatedly described as wanderings and labours in the surviving *Epic*; the *Odyssey* is a story of wanderings and labours; Herakles' journeys to Spain and/or India, and Dionysos/Liber's journey to India (well-recognised as a duplication of the Herculean travels) can be found described as a 'wandering'; Dionysos' duplicate, Osiris, is described by Plutarch as having made similar wanderings, which are contrasted with his labours; when Alexander's troops mutiny, they complain of their wandering, and are reminded of the parallel between their victories and the labours of both Dionysos and Alexander's forefather Herakles. Book I of the *Aeneid* furthermore ends a few lines after the passage discussed above, with (lines 755—6) a repeated reference to Aeneas' wanderings since the fall of Troy. Livy similarly speaks of the 'interminable wanderings' [*ab inmenso prope errore*] of Aeneas' men, and contrasts the 'labours' [*operum*] of Aeneas' later battle with Turnus and the Etruscans. When Virgil later describes this same battle, he breaks his narrative to tell us the story of Herakles' combat with Cacus, and the parallel between Aeneas and Herakles is obvious.[32] One must suspect that the wander/labour contrast in 1.742 is simply part of a larger attempt to set the story of the founding of Rome against a background which echoes with overtones of the 'paragon of labours and wandering'. One of Herakles' traditional roles was in fact the pacification of a territory in preparation for the founding of a city.[33]

Virgil's song is sung furthermore by a lyre-carrying associate of the Muses who (we are told) had learned the astronomy of his song from Atlas, and lines 740—2 of the poem end with the words 'Iopas' (the bard), 'Atlas' and 'labores' respectively, strongly suggesting that Virgil sees these three as linked. But Herakles was also an evident associate of the Muses in 1c BC Rome — the coins noted above had been minted in Virgil's lifetime — and Atlas was similarly the traditional teacher of Herakles. Indeed, one of our sources for this fact is Servius' commentary on these lines of Virgil, and this indicates that Servius certainly recognised a reference to Herakles in the passage.

The myth that Atlas held up the sky, Servius and others tell us, was often interpreted as meaning that Atlas had a thorough understanding of the structure of the heavens, and the fact that Herakles temporarily took his burden from him meant that Herakles acquired this understanding from him. He then passed it on to the Greek astronomers. As Diodorus of Sicily explains:[34]

> They also say that [Atlas] perfected the science of astrology and was the first to publish to mankind the doctrine of the sphere; and it was for this reason that the idea was held that the entire heavens were supported upon the shoulders of Atlas, the myth darkly hinting in this way at his discovery and description of the sphere ... Similarly in the case of Heracles, when he had brought to the Greeks the doctrine of the sphere, he gained great fame, as if he had taken over the burden of the firmament which Atlas had borne, since men intimated in his enigmatic way what had actually taken place.

These various ancient connections between Herakles and astronomy were well-known in the Renaissance, and historians already acknowledge that they lie behind some of the emblematic uses of Herakles in early modern decoration.[35] Most sixteenth and seventeenth century mythographers note the belief that Herakles is the sun, and many also explain that his labours represent the motion around the zodiac. Thus Alexander Ross writes:[36]

COPERNICUS, APOLLO, AND HERAKLES 15

By *Hercules* some understand the Sun, who is ... the glory of the air, which is then glorious, when by the Sun beams it is illuminate. His 12 labors are the 12 signes in the Zodiack, which, every year, he passeth thorow: he is the son of Jupiter and Alcmene [whose name] signifieth strength or power, because God by his Almighty power created the Sun, and gives power to the Sun to overcome all the oppositions of Cloudes, Mists, Vapors, which [Hero] (or Juno) the air cast before him to obscure his light. Hebe the goddess of Youth is married to him, because when he returns to us in the Spring, he reneweth all things, and makes the world as it were youthful again. Geryon, whome Hercules overcame, is the Winter, which the Sun masters.

Equally commonly cited is the belief that Herakles was an expert astronomer, who studied under Atlas. The Venice 1544 edition of Virgil, for example, contains a number of ancient commentaries which reproduce these beliefs; and Fraunce's *Third Part of . . . Yuychurch*, the '[m]ost important of the English 16th-century mythographies', explains Herakles' labour in the garden of the Hesperides as an astronomical myth. The 'daughters of *'Hesperus'*, we are told,[37]

are the starres: their garden is in the weast, wherein grow golden apples: for such is the nature of the starres, to glister like gold, and seem round in show like apples. They grow in the weast, because the stars neuer appeare, but when the sunne setteth, and that is in the weast: for, all the day long they are obscured by the surpassing light of the sunne. The never sleeping Dragon, that watcheth these apples & keepeth the garden, is the [celestial] circle called Signifer. Hercules brought these into Greece, that is, he brought Astrologie into his country. So was he, for the same cause, fayned to beare the heavens on his shoulders, whilst Atlas rested himself: because he learned Astrologie of Atlas: who is therefore sayd to holde up the heavens, because he continually observed the motions of the heavens ... The Pleīdes and Hyades be called his daughters, because he first noted their course, and observed their operation.

So Mossakowski's obviously attractive identification of the seal as a representation of Apollo can be readily paralleled by a similarly compelling Herculean interpretation. Clearly neither of these interpretations is in itself satisfactory — Apollo and Herakles have, by Copernicus' time, such similar traditions attached to them, that we cannot use knowledge of these traditions to identify the figure in the seal as one of these gods rather than the other. Indeed, Apollo would seem to have become an emblem of Herakles, and Herakles an emblem of Apollo, so that to identify the seal as one of these gods is to identify it *a fortiori* as the other. Furthermore, some of the evidence introduced above would seem to suggest that it might be possible to develop a plausible argument that the seal also represents Mercury, a god we already know to have been used as a seal-image by other members of the Varmian canonry.[38] This suggests that it is better to think of the seal as representing what Apollo and Herakles (and perhaps also Mercury?) have in common.

The most obvious thing that Apollo and Herakles have in common is that both are monster-fighters, for Apollo had to kill the serpent Python to establish his sanctuary at Delphi. Hermetic mythology too includes a battle with many-eyed Argos. Such combats are in fact a common feature of classical mythology, and detailed scrutiny of the accounts of them which survive makes it clear that the various stories are extremely closely linked, so that the superficially different heros are to a large extent duplicates of each other.[39] Many of the stories furthermore are distinctly musical in character, and they are repeatedly presented as stories of the creation or maintenance of harmony — via the defeat of some monster who threatens the unity of the cosmos or its analogues. Hence arises the particular link needed here between Herakles and the harmony of the universe.[40]

More generally too, our seal emerges as an emblem of the common hero of this whole family of myths. So when Copernicus, in his dedication of *De Revolutionibus* to the Pope, expresses his dissatisfaction with Ptolemaic astronomy by characterising prevailing accounts of the planetary motions as monstrous, he is expresisng his desire to reveal the harmony of the cosmos via much the same metaphor as that contained in his choice of a musician-hero as personal seal.[41] '[T]hose who have devised eccentric circles ... have not been able to discover or deduce from them the chief thing, that is the form of the universe, and the clear symmetry of its parts. They are just like someone including in a picture hands, feet, head, and other limbs from different places, well painted indeed, but not modelled from the same body, and not in the least matching each other, so that a monster would be produced from them rather than a man.'

NOTES

[1] Stanislaw Mossakowski. 'The Symbolic Meaning of Copernicus' Seal.' *Journal of the History of Ideas*, **34**: 451—460 (1973).

[2] Mossakowski, 'Symbolic Meaning', *op. cit.* note 1, 454—7. There are various spellings of Gaforus' name and book-title in circulation: I follow the *British Library Catalogue*. The inscription on the banner comes from *Nomina Musarum* (line 9) in the

appendix to Ausonius, transl. H. White, Loeb series, II, London (1961), pp. 280—1. The poem was apparently thought to be Virgilian in the Renaissance: see Mossakowski, *loc. cit.* and Edgar Wind, *Pagan mysteries in the Renaissance*, rev. edn. New York (1968), pp. 267—8. The next line of *Nomina Musarum* describes the sun (Apollo-Phoebus) as 'residing in the middle' (*in medio residens*).

[3] For a detailed survey of the myths of the various Herakles, see Robert Graves, *The Greek Myths*, 2 Vols. Harmondsworth (1966), *passim*, esp. II, pp. 84—206.

[4] A good seventeenth century examples is provided in the *Galerie d'Apollon* at the Louvre: one of the main ceiling paintings here depicts Apollo with his lyre, accompanied by his sister Artemis. The immediately neighbouring panel contains a painting of the zodiac, and another depicts Apollo's victory over Python, while the whole room is full of references to astronomy, music, the Muses, and Apollo. A particular clear (but rather late) example is the ceiling painting (by Lemoine, dated 1733—6) in the Hercules Drawing Room at Versailles. Though the main theme here is the Apotheosis of the hero, the harmony of the universe is overtly referred to by the inclusion of Apollo and the Muses: one of the latter (Euterpe?) plays a lyre, next to another (Urania?) who holds a celestial globe. There are also many references to Herakles in the original seventeenth century portions of the palace. Two good earlier examples of this imagery, neither attached to a strictly royal palace, are provided by the Piazza della Signoria in Florence, and the main entrance stairway to the Doge's palace in Venice. The latter is flanked by a pair of Herculean statues, one holding the celestial globe, and the other defeating the Lernaean hydra. The *Piazza* is dominated by numerous statues of heros who represent the myth-complex discussed in the body of the present paper: Herakles (and Cacus), David, Perseus, Judith (and Holofernes), and a celestial Neptune — all paralleling an equestrian Cosimo I de' Medici — while the whole cycle of Herakles' labours is represented inside the *Palazzo Vecchio*. For some further literature see n. 6, on pp. 97—8 of Keith Hutchison, 'Towards a Political Iconology of the Copernican Revolution', on pp. 95—141 of Patrick Curry (ed.), *Astrology, Science and Society*, Woodbridge, Suffolk (1987), and Mario Biagioli, 'Galileo the Emblem-Maker,' *Isis*, **81**: 230—58 (1990).

[5] 'Towards a Political Iconology of the Copernican Revolution', *op. cit.* note 4, esp. pp. 109—112 nn. 32, 33.

[6] Mossakowski, 'Symbolic meaning', *op. cit.* note 1, p. 451 nn. 2—3, citing Adolf Furtwängler, *Die Antiken Gemmen: Geschichte der Steinschneidekunst im klassischen Altertum*, 3 Vols, Amsterdam (1964—5; 1900), I, pl. XXXIX—4, II, p. 186, III, p. 348.

[7] See: Furtwängler, *Antiken Gemmen*, *loc. cit.* note 6, plus I, pl. XLIV—60; Georg Lippold, *Gemmen und Kameen des Altertums und der Neuzeit*, Stuttgart (n.d.), pl. VIII, pl. IX; Gisela Richter, *Engraved Gems of the Greeks Etruscans and Romans*, part II: *Engraved Gems of the Romans*, London (1971), figs 71, 72, 73, 74, 78, 251—3, 678 (perhaps also 727, 727a, 728, 729 which may not however be ancient.)

[8] Furtwängler, *Antiken Gemmen*, *op. cit.* note 6, I, pl. XXVII—13, 14, pl. XXXIV—25, pl. LVII—10, described II, pp. 134—5, 167, 260; Lippold, *Gemmen und Kameen*, *op. cit.* note 7, pl. XXXIX (identical with Furtwängler pl. LVII—10); Richter, *Gems of the Romans*, *op. cit.* note 7, figs 279, 692 (again identical with Furtwängler pl. LVII—10). Kathi Meyer-Baer, *Music of the spheres and the Dance of Death: Studies in Musical Iconology*, Princeton (1970), p. 268 claims that there are 'several works' depicting

Herakles as a musician, but only gives a precise reference to Furtwängler, pl. LVII—10, and its repetition in Lippold.

[9] For Apollo, see Furtwängler, *Antiken Gemmen, op. cit.* note 6, pl. XXXIX—4; Richter, *Gems of the Romans, op. cit.*, note 7, fig. 72; for Herakles, Furtwängler pl. XXXIV—25 and LVII—10. For the coins, see notes 11 and 23, and, for Renaissance familiarity with them, notes 20 and 23.

[10] Ovid, *Fasti*, VI. 797—812, transl. with notes etc. by J. G. Frazer, Loeb series, London (1976), pp. 382—3. Note that Ovid's 'Alcides' is Herakles: cf. *ibid.*, I. 575 (= Loeb, pp. 42—3) and Frazer's notes b and c, Loeb p. 382. Copernicus cites this work in his latin translation of Theophylactus Simocatta: see *Nicholas Copernicus, Minor Works*, ed. P. Czartoryski, transl. E. Rosen and E. Hilfstein, London (1985), pp. 29, 51 (first note 6).

[11] For the temple in Rome, see: A. Pauly, G. Wissowa and W. Kroll, *Real-Encyclopädie der Classischen Altertumswissenschaft*, Vols 1+, Stuttgart (1894+), VIII, pp. 574—6; R. Peter in W. H. Roscher, *Ausführliches Lexikon der Griechischen und Römischen Mythologie*, Vol I, part 2, Leipzig (1886—1890), cols 2970—6; H. A. Grueber, *Coins of the Roman Republic in the British Museum*, 3 Vols, London (1970; 1910), I, p. 442n. I have not been able to consult the principal literary source for the founding of this temple, Eumenius (the third century rhetorician), *Pro Rest. Schol.*, 7. The works just cited note the important evidence for this temple provided by coins (as in Fig. 4), and Pauly *et al.*, indicate that it is also documented via a fragment of a townplan. For the passages cited, see: Ovid, *Art of Love*, III. 165—8, transl. by J. Mozley, Loeb series, II, London (1969), pp. 128—9; Suetonius, *Divi Augusti Vita*, 29; Macrobius, *Saturnalia*, I.xii.16, transl. P. Davies, New York (1969), p. 87; Plutarch, *Roman Questions*, 59 = *Moralia*, 278D—E, transl. F. Babbitt, Loeb series IV, London (1936), pp. 94—5.

[12] Xenophon, *Memorabilia*, 2.1.21—34 (= Loeb, pp. 94—103); Dio Chrysostom, *First discourse, On Kingship*, 58—84 (= Loeb I, pp. 31—47), *Eighth Discourse, On Virtue*, esp. 27—36 (= Loeb I, pp. 390—99), *Thirty-First Discourse*, 16 (= Loeb III, pp. 20—1). Cf. G. Karl Galinsky, *The Heracles Theme: The Adaptions of the Hero in Literature* . . . , Oxford (1972), passim., but especially pp. 35—6, 39n17, 101—3, 162, 198ff; R. Hoistad, *Cynic Hero and Cynic King*, Lund (1948), pp. 150ff; Erwin Panofsky, *Hercules am Scheideweg*, Leipzig (1930), pp. 43—52, 150—66.

[13] Plato, *Republic* VII, 528E—531D (= Loeb II, pp. 178—195); Aristotle, *Politics*, VIII.ii.3 (1337b20—35), VIII.iv.3—vii.11 (1339a10—1342b35) (= Loeb, pp. 638—9, 648—75); Plutarch, *Alcibiades*, II.4—6 (= Loeb *Lives* IV, pp. 6—89); *Oxford Classical Dictionary*, edited N. Hammond and H. Scullard, 2nd edition, Oxford (1970), p. 705 ('Music'); *The New Grove Dictionary of Music and Musicians*, ed. S. Sadie, I, London (1980), pp. 699—702 ('Aulos'); Jamie Kassler, 'Apollo and Dionysos: Music Theory and the Western Tradition of Epistemology', *Music and Civilization: Essays in Honor of Paul Henry Long*, E. Strainchamps and M. Maniates (eds.), London (1985), pp. 457—8; John Hollander, *The Untuning of the Sky: Ideas of Music in English Poetry 1500—1700*, Princeton (1961), p. 35; W. K. C. Guthrie, *The Greeks and their Gods*, London (1977), pp. 48, 148, 157, 314—5; Stephen Kolsky, 'Images of Isabella d'Este', *Italian Studies*, **39**, 47—62, p. 53 (1984); Meyer-Baer, *Music of the Spheres, op. cit.*, note 8, p. 205.

[14] Pausanius, *Description of Greece*, IX.29.6—9 (= Loeb IV, pp. 296—99); Diordorus of Sicily, *History*, III.67.1—3 (= Loeb II, pp. 304—7); Apollodorus, *The Library*, I.iii.2, II.iv.9, transl. J. G. Frazer, Loeb series, 2 Vols, London (1967—1970), I, pp. 16—17, 174—7; Frazer's note 5 to Apollodorus, *op. cit.*, Loeb I, p. 16; Joseph Fontenrose, *Python: A Study of Delphic Myth and its Origins*, Berkeley (1959), pp. 111—2. For examples of the paralleling of the bow (Apollo's canonical weapon) with the lyre, see: the juxtaposition of the two images of Apollo in Callimachus, *Hymn to Apollo*, lines 19, 33, 43—4, 90—110 (= Loeb Callimachus and Lycophron and Aratus, pp. 50—3, 56—9), as discussed Frederick Williams, *Callimachus Hymn to Apollo: A Commentary*, Oxford (1978), pp. 29—30, 40; Heraclitus' illustration of the doctrine of the harmony of opposites via the metaphor of the lute and bow, as quoted Plutarch, *Moralia*, 369B (= Loeb V, pp. 108—9), and as discussed W. K. C. Guthrie, *A History of Greek Philosophy*, 5 Vols, Cambridge (1962—78), I, pp. 439—440.

[15] Fontenrose, *Python, op. cit.* note 14, pp. 339—40; Jocelyn Small, *Cacus and Marsyas in Etruscan and Roman Legend*, Princeton (1982), *passim*. but esp. pp. xiv, 4 and Figs 1, 3, 4, 6; Virgil, *Aeneid*, VIII.305, transl. H. Rushton Fairclough, Loeb series, 2 Vols, London (1967—1969), II, pp. 80—1.

[16] Diodorus of Sicily, *History*, III.67 (= Loeb II, pp. 304—309); Plutarch, *Moralia*, 278D—E (= Loeb IV, pp. 94—5). Compare the close connection between the linguistic and musical skills of the Muses in Hesiod, *Theogony*, lines 75—103 (= Loeb, pp. 82—5).

[17] For characteristics of Mercury, see: Graves, *Greek Myths, op. cit.* note 3, I, pp. 63—7; *Oxford Classical Dictionary, op. cit.* note 13, pp. 502—3 ('Hermes (1)'); Diodorus of Sicily, *History*, V.75 (= Loeb III, pp. 300—301); Plutarch, *Moralia*, 352A (= Loeb V, pp. 10—11); Hyginus, *Poetica Astronomica*, II.42; Anthony Pomey, *The Pantheon*, transl. J. A. B., London (1694); reprinted New York (1976); first Latin edn (1653), pp. 60—4; Meyer-Baer, *Music of the Spheres, op. cit.*, note 8, p. 288.

[18] Lucian, *Herakles*, as transl. H. W. and F. G. Fowler, *The Works of Lucian of Samosata*, III, Oxford (1905), pp. 256—9. It should however be noted that Lucian attributes the view of Herakles as eloquent to the Gauls, not to the Greeks and Romans. But compare Plutarch, *Moralia*, 356A—B (= Loeb V, pp. 34—5) on Dionysos. See also Graves, *Greek Myths, op. cit.* note 3, II, pp. 92, 93n7, 142—3n3, 212—3.

[19] Lilio Giraldi (1479—1552), *De Deis Gentium*, Basel (1548); reprinted New York (1976), p. 456. This book was dedicated to a modern Herakles, Ercolo II, the fourth d'Este duke of Giraldi's home-town Ferrara — where Copernicus himself had lived some 45 years before the appearance of his near-contemporary's book. It appears though not to be known when Copernicus started using his seal: see Mossakowski, 'Symbolic Meaning', *op. cit.* note 1, p. 453. Gerardus Vossius, *De Theologia Gentili*, Amsterdam (1641); reprinted in 3 Vols, New York (1976), I, second p. 87; Andrea Alciati [and C. Mignault], *Emblemata cum Commentariis*, Padua (1621); reprinted New York (1976), pp. 129, 599, 775. The quoted descriptions of these books comes from Stephen Orgel's introductory notes to the reprints used.

[20] Vicenzo Cartari [and Lorenzo Pignoria, annotator], *Imagini delli Dei de gl'Antichi*, Venice (1647); reprinted Graz (1963), pp. 180—1, 320—1 (my translation). The first edition of this book came out in 1556, but was unillustrated, and did not have the

annotations by Pignoria. The next (?) edition (1571; reprinted New York (1976), dedicated like Giraldi's *De Deis Gentium* (*op. cit.* note 19) to the d'Este family of Ferrara) had engravings by Bolognino Zaltieri, and included the Lucian image on pp. 340—1. It is not in the unillustrated 1599 English truncated translation by Richard Lynche, *The Fountaine of Ancient Fiction*, (reprinted New York, 1976), but the eloquence of Herakles is mentioned (again within the discussion of Mercury, at Si, unpaginated). The Pignoria notes were first (?) attached to the Padua, 1615 edition, *Le Vere e Nove Imagini de gli Dei delli Antichi*, where the 'Copernican' image is included on pp. 545—6. The quoted description of Cartari's book comes from Stephen Orgel's introductory note to the New York (1976) reprint of the 1571 edition. For the Renaissance's especial recognition of Herakles' eloquence, see: Galinsky, *Herakles Theme, op. cit.* note 12, pp. 222—4; Giraldi, *De Deis Gentium, op. cit.*, note 19, p. 455; Bocchi, *Symbolicarum Quaestionum*, Bologna (1574); reprinted New York (1979), pp. 92—3; Abraham Fraunce, *The Third Part of . . . Yuychurch*, London (1599); reprinted New York (1976), p. 49; Alciati, *Emblemata, op. cit.*, note 19, pp. 599, 753; Alexander Ross, *Mystagogus Poeticus, or The Muses Interpreter*, 2nd edn, London (1648); reprinted New York (1976), pp. 170, 187; Vossius, *De Theologia Gentili, op. cit.*, note 19, I, pp. 266, 382.

[21] Cartari, *Imagine, op. cit.*, note 20, 1647 edn, p. 320—1. Hollander, *Untuning of the Sky, op. cit.*, note 13, pp. 194—206, and illustrations (unnumbered) following p. 242.

[22] Albrecht Altdorfer (1480—1531), 'Hercules and a Muse' (B28 = Bartsch VIII, p. 51), reproduced in my Fig. 6 from Adam von Bartsch, *The Illustrated Bartsch*, Vols 1+, New York (1978+), XIV, p. 36. Master of the Die (= Le Maître au dé), 'Envy Driven from the Temple of the Muses' (B17 = Bartsch XV, p. 195), reproduced in *Illustrated Bartsch*, XXIX, p. 174. Enea Vico, (1523—1567), engraving 'Man Playing a Lyre', (B121 = Bartsch XV, p. 322), reproduced in my Fig. 7, from *Illustrated Bartsch*, XXX, p. 95: this is easily recognized as Herakles because of the lion skin over the shoulder. Laurentius Beger, *Hercules Ethnicorum*, Berlin (1705), pp. 31—2: note that one of Beger's figures is labelled with the Apolline epithet 'Herculi Musarum *Pythius*' (my emphasis). For a near-contemporary (1694) application of the epithet 'Pythius' to Apollo, see Pomey, *Pantheon, op. cit.* note 17, p. 45.

[23] For the Roman coins, see: Pauly, Wissowa & Kroll, *Encyclopädie, op. cit.* note 11, VIII, p. 576; Roscher, *Lexikon, op. cit.* note 11, Vol. I, col. 2972; Grueber, *Coins, op. cit.* note 11, Vol. I, pp. 251 (n. 2), 441—2 (nn. 1, 2), 441—6 (coins 3602—3632, esp. 3602—5), 450—1 (n. 2), Vol. III, pl. xlv—13. For the Greek medal, see Guillaume du Choul, *Discours de la Religion des Anciens Romains . . . Illustré*, Lyon (1556); reprinted New York (1976), p. 180. The quoted description of this book comes from S. Orgel's introductory note to the reprint.

[24] Macrobius, *Saturnalia*, I.xx.6—12, *op. cit.*, note 11, p. 138—9; Nonnos, *Dionysiaca*, XL. 369—410, transl. W. Rouse with notes etc. by H. Rose and L. Lind, Loeb series, 3 Vols, London (1940), III, pp. 180—3. Cf. Note by Rose, *ibid.*, pp. 194—5. For Renaissance citations, see note 36.

[25] Graves, *Greek Myths, loc. cit.* note 3; Jonathan Brown and J. Elliot, *A Palace for a King: The Buen Retiro and the Court of Philip IV*, New Haven (1980), pp. 156—161, 272—3; Hildegard Utz, 'The *Labors of Hercules* and Other Works by Vicenzo de' Rossi', *Art Bulletin*, 53, 344—66, esp. 356—60 (1971). Graves cites no precise passages for this point, and scrutiny of the sources cited by Brown and Utz, shows the former to be completely dependent on the latter, and the latter completely dependent

on Cartari, *Imagini, op. cit.* note 20, (locations as cited below, note 36). Cartari's ancient source is Macrobius, but again no precise passage is cited. When Coluccio Salutati similarly cites Macrobius, in his 1406 *De laboribus Hercules*, III.15 (ed. B. L. Ullman, Zurich (1951), p. 168), his modern (and seemingly thorough) editor identifies the passage as that quoted above (p. 10): there is no reference there to the belief that the labours correspond to the signs of the zodiac. This suggest that Macrobius does not in fact discuss the issue. Similarly, when George Sandys discusses the issue in his *Ovids Metamorphosis Englished, Mythologized, and Represented in Figures* ... , Oxford (1632); reprinted New York (1976), p. 325, he cites Macrobius for an identification with sun, but cites (p. 329) Porphry (unspecified location) for the identification with the signs of the zodiac.

[26] Plutarch, *Moralia*, 367E (= Loeb V, pp. 100—101); *The Epic of Gilgamesh*, transl. N. K. Sandars, rev. edn, Harmondsworth (1972), pp. 24, 36—7, 61, 71—2, 79, 89—92, 96—99; Fontenrose, *Python, op. cit.* note 14, pp. 167—75; Graves, *Greek Myths, loc. cit.* note 27.

[27] Graves, *Greek Myths, op. cit.* note 3, II, pp. 87(§i), 89(n. 2), 103(n. 3), 105(§h), 106—7(nn. 1, 3), 108—9(§g and n. 2), 114(§g), 151(n. 3). The lion, the bull, the scorpion, the water-snake (*sic*), the crab, and the centaur are the zodiacal constellations which Graves explicitly notes to be connected with Herakles' labours. Cf. C. Kerényi, *The Heroes of the Greeks*, London (1974), pp. 143, 145, 198.

[28] Ovid, *Metamorphoses*, II.52—89, 385—7 (= Loeb I, pp. 64—7, 86—7), as transl. M. Innes, Harmondsworth: Penguin (1982), and F. Miller, Loeb series, I, London (1956), (though I have replaced Miller's original 'toils' by 'labour'). Virgil, *Aeneid*, I.742. Ovid (it will be observed) blends the sun's *daily* motion through the sky, with its *annual* motion around the ecliptic: I presume this to be a distinction often overlooked by 'amateurs'. For other descriptions of the sun's activity as a 'labour', see Ovid, *op. cit.*, VI. 486, Loeb I, p. 322—3, plus passages cited in James Henry's commentary on the line of Virgil just cited, in his *Aeneidea, or Critical, Exegetical and Aesthetical Remarks on the Aeneis*, Vol. I, London (1873), pp. 852—3.

[29] For the 'eclipse' interpretation, see: John Conington and Henry Nettleship, commentary on *Aeneid*, I.742 in *The Works of Virgil*, 5th edn, 3 Vols, London (1898); reprinted Hildesheim (1963), II, p. 83, and on *Georgics*, II.478, *ibid.*, I, p. 277; R. G. Austin, commentary, in Virgil, *Aeneid*, Oxford (1971), I, p. 224; T. E. Page, notes to *The Aeneid of Virgil: Books I—VI*, London (1938), p. 205; Charlton Lewis and Charles Short, *A Latin Dictionary*, Oxford (1966), p. 1024 (*labor*); and (for the tradition behind this interpretation), Virgil [and collected commentators], *Opera*, Venice (1544); reprinted New York (1976), I, p. 99 (commentaries by Servius, Probus and Mancinellus on *Georgics*, II.478), p. 199 (commentary by Ascensius on *Aeneid*, I.742).

[30] Lucan, *De Bello Civili*, I.i.90; Cicero, *De Natura Deorum*, II.xl—xli, esp. p. 104, transl. H. Rackham, Loeb series, London (1961), pp. 220—223. The *labuntur* occurs in a phrase which Cicero has taken from his own transl. of Aratus, *Phaenomena*: see H. Rackham's note a, *ibid.* (Loeb) p. 222, and G. Mair's note e, *Callimachus, Lycophron and Aratus*, transl. A. and G. Mair, Loeb series, London (1921), p. 381. The notion does not seem to be present in the original Greek, but it is also added (?) in line 18 (*indefessa*, 'never tiring') of the Germanicus (?) translation: see *The Aratus Ascribed to Germanicus Caesar*, ed. with notes etc. by D. Gain, London (1976), pp. 20, 53. Manilius, *Astronomica*, III.146, pp. 160—164, transl. with notes etc. by G. P. Goold, Loeb series, London (1977). For the dating of Manilius, see Goold's introduction, *ibid.*,

pp. xi—xii. See also Goold's remarks on the term *athlum/athla, ibid.*, pp. lxii—lviv, 174 (note c), and Lewis and Short, *Dictionary, op. cit.* note 29, p. 188 (*athlon*); Henry Liddell and Robert Scott, *A Greek-English Lexicon*, rev. etc. by H. Jones and R. McKenzie, Oxford (1968), pp. 32—3. For the myth of the Herculean founding of the Olympic Games, see Pindar, *Olympian Odes* X (= Loeb, pp. 112—5), Diodorus of Sicily, *History*, III.74 (= Loeb II, pp. 330—1).

[31] *The Aeneid of Thomas Phaer and Thomas Twyne: A critical edition...*, ed. S. Lally, New York (1987), p. 26: this portion of the translation is dated 1555. For other examples of the zodiacal motion interpretation, see: Servius and Donatus, in Virgil, *Opera*, (1544), *op. cit.*, note 29, pp. 189—9; Jacobus Pontanus, *Symbolarum Libri XVII Virgilii*, 3 Vols, Augsburg (1599); reprinted New York (1976), I. cols 424—5 (on *Georgics*, II. 478), II, cols 813—4 (on *Aeneid*, I. 742) (who interprets the word as either the motion of the sun, or its eclipse); comparison between descriptions of motions of moon (p. 42b) and sun (p. 33b), Fraunce, *Third part, op. cit.*, note 20; Henry, *Aeneidea, op. cit.*, note 28, I, p. 852.

[32] Livy, I.i.1—I.ii.6 (= Loeb I, pp. 8—13); Horace, *Odes*, III.iii.9, (= Loeb, pp. 178—9); Dio Chrysostom, *Eighth Discourse, On Virtue*, 27—9 (= Loeb I, pp. 390—3); *Epic of Gilgamesh, op. cit.* note 26, pp. 61, 91, 97, 101—6, 117; Plutarch, *Moralia*, 341F—342A, 361D (= Loeb IV, pp. 468—9, V, pp. 66—7), *Alexander*, XLI.1 (= Loeb *Lives* VII, pp. 344—5); Arrian, *Anabasis*, V.4, V.26, 29, VII.1, VII.9—10 (= Loeb II, pp. 10—11, 84—91, 96—97, 206—7, 232—3); Quintus Curtius Rufus, IX.ii.29, IX.iii.14 (= Loeb II, pp. 384—5, 390—1, where Pater Liber is Dionysos); Homer, *Odyssey*, (e.g., I.2, = Loeb pp. 2—3); Apollodorus, *Epitome*, VI.29—VII.1 (= Loeb II, pp. 278—9); Nonnos, *Dionysiaca*, IV.270, 287, 293 (= Loeb I, pp. 152—5); Lynche, *Fountaine, op. cit.*, note 20, pp. Tb—Tijb.

[33] Diodorus of Sicily, *History*, I.13—20, III.64, 72—4, V.76 (= Loeb I, pp. 44—65, II, pp. 296—7, 322—333, III, pp. 302—3); Plutarch, *Theseus*, VI, XXIV—XXV (= Loeb *Lives* I, pp. 12—17, 50—57). Cf. Plutarch, *Moralia*, 332A—B, 356A—B (= Loeb IV, pp. 412—3, V, pp. 34—5); Du Choul, *Discours, op. cit.* note 23, pp. 182—4; Galinsky, *Herakles Theme, op. cit.* note 12, p. 132 (and p. 12, from which the quoted description of Herakles is taken.)

[34] Diodorus of Sicily, *History*, III.60, IV.27, transl. H. C. Oldfather, Loeb series, vol.II, London (1961), pp. 278—9, 430—1; Servius, in the 1544 Venice Virgil, *loc. cit.*, note 29; Augustine, *City of God*, XVIII.8, (= Loeb V, pp. 386—9). Compare: Graves, *Greek Myths, op. cit.* note 3, II, 150—1n1. For Renaissance examples, see below note 37.

[35] See the discussions by Brown and Elliot and by Utz cited above, note 25.

[36] Ross, *Mystagogus Poeticus, op. cit.* note 20, p. 168. Cf. Salutati, *De laboribus Hercules, op. cit.* note 25, p. 168; Natalis Comes, *Mythologie*, transl. J. Baudouin, Paris (1627); reprinted in 2 Vols, New York (1976), (II), pp. 708—9; Sandys, *Ovid, op. cit.*, note 25, pp. 322, 325 (misnumbered as 335)—7, 329; Alciati, *Emblemata, op. cit.* note 19, p. 599; Vossius, *De Theologia Gentili, op. cit.*, note 19, I, pp. 382—5; Cartari, *Imagini, op. cit.*, note 20, 1571 edn, p. 450, 1647 edn, p. 184. Philostratus, *Les Images*, transl. B. de Vigenère, Paris (1614); reprinted New York (1976), p. 464.

[37] Virgil (and commentators), *Opera* (1544), *op. cit.*, note 29, p. 199 (Servius and Ascensius); Fraunce, *Third part, op. cit.*, note 20, p. 46b—47a (misnumbered as 49). The quoted description of Fraunce's book comes from S. Orgel's introduction to the reprint edition. Cf.: de Vigenère, in Philostratus, *Les Images, op. cit.*, note 36, p. 466;

Vossius, *De Theologia Gentili op. cit.*, note 19, frontispiece, and I, p. 384; Giraldi, *De Deis Gentium op. cit.*, note 19, p. 457; Giovanni Boccaccio, *Genealogiae*, Venice (1494); reprinted New York (1976), p. 97; Bocchi, *Symbolicarum Quaestionum, op. cit.*, note 20, pp. 236—7; Natalis Comes, *Mythologiae*, Venice (1567); reprinted New York (1976), p. 105, *Mythologie op. cit.*, note 36, II, pp. 312—3 (a passage which cites the *Aeneid* line I.742 discussed above); Sandys, *Ovid, op. cit.*, note 25, pp. 167, 327; Salutati, *De laboribus Hercules, op. cit.* note 25, p. 309; Ross, *Mystagogus Poeticus, op. cit.*, note 20, pp. 37, 169, 175.

[38] Mossakowski, 'Symbolic meaning', *op. cit.*, note 1, p. 453. Mossakowski here says (n. 8) that the Mercury image was to have been published in vol. III of the Copernicus *Opera Omnia*. This presumably refers to *Nicholas Copernicus, Minor Works, op. cit.*, note 10, but I have not been able to find it there.

[39] Fontenrose, *Python, op. cit.*, note 14, is a sustained argument for the unity of the myth-variants in question, with particular emphasis on the relationship between Apollo and Herakles, and I rely greatly on his account here. Compare the similar arguments and presumptions in: Mircea Eliade, *Cosmos and History: The Myth of the Eternal Return*, transl. W. Trask, New York (1985), pp. 37—9; Francis Cornford, *Principium Sapientiae: The Origins of Greek Philosophical Thought*, Cambridge (1952), pp. 226—56. Note that even the dissident view of J. G. Griffiths, *The Conflict of Horus and Seth: From Egyptian and Classical Sources*, Liverpool (1960), pp. 128—30 accepts much of what the proponents of unity claim.

[40] Fontenrose, *Python, op. cit.*, note 14, has a whole chapter on this theme (pp. 217—73) but he tends to take the issues which concern us here for granted, and I plan to prepare a further study of this issue as a sequel to the present paper. For Fontenrose does not provide systematic evidence for the links between combat and cosmic harmony, beyond those implied in the occurrence of monster-combat in creation myths, where the harmony of the universe is established. The sort of primary evidence used to establish my more general interpretation is that provided in Plutarch's discussion of the myth of Isis and Osiris, *Moralia*, 351C—384C (= Loeb V, pp. 6—191), and books I and II of Nonnos, *Dionysiaca*, (= Loeb I, pp. 2—97), the most important cosmic passages being I.163—258, II.163—435, 565—619, 650—678. Cf. Jean-Pierre Vevnant, *The Origins of Greek Thought*, Ithaca, N.Y. (1982), pp. 108—118.

[41] Nicolas Copernicus, Letter to the Pope, in *De Revolutionibus Orbium Coelestium*, Nuremberg (1543), transl. by A. M. Duncan as *On the Revolutions of the Heavenly Spheres*, New York (1976), p. 25. Compare: Bocchi, *Symbolicarum Quaestionum*, (1574), *op. cit.*, note 20, pp. 236—7; extract from the preface to Thomas Digges, *Alae sue scalae mathematicae*, London (1573), as transl. on p. 347 of J. Dreyer, *A History of Astronomy from Thales to Kepler*, 2nd edition, revised W. H. Stahl, New York (1953); Galileo Galilei, *Dialogue Concerning the Two Chief World Systems — Ptolemaic and Copernican*, transl. S. Drake, Berkeley (1962), pp. 341, 366. Copernicus' immediate source appears to be the opening lines of Horace's, *Art of Poetry*, (= Loeb, pp. 450—1): see *Nicholas Copernicus On the Revolutions*, ed. J. Dobrzycki, transl. with commentary by E. Rosen, London (1978), pp. 338 (note to 3:37), 341 (note to 4:25). But the image was a common one in ancient literature: see Fontenrose, *Python, op. cit.*, note 14, pp. 152, 207, 243, 308, and Apollonius of Rhodes, *Argonautica*, IV.673—681 (= Loeb, pp. 338—41).

JOHN SUTTON*

RELIGION AND THE FAILURES OF DETERMINISM

> Fate's a spaniel,
> We cannot beat it from us.
>
> John Webster, *The White Devil*[1]

INTRODUCTION

To trace a path from Pico della Mirandola's Renaissance man to the Jacobean malcontents of Marston or Webster is to document not an inflation of hopes for dominion over the natural world, but rather a loss of confidence in the possibility of control over even human affairs. 'For I am going into a wilderness, /Where I shall find nor path, nor friendly clew/To be my guide'.[2] The bleak consequences of this lack of direction, leaving traces through into the Restoration period in England, are particularly evident in the free will debate: of Milton's angels,

> Others apart sat on a hill retired,
> In thoughts more elevate, and reasoned high
> Of Providence, Foreknowledge, Will, and Fate-
> Fixed fate, free will, foreknowledge absolute-
> And found no end, in wandering mazes lost.[3]

For Pico, man is of intermediary status, unique among earthly creatures in being linked to the divine mind. This is the source of his glory, an optimistic encouragement to try to ascend the chain of being. But, a generation after Pico's *Oration on the Dignity of Man*, the same reflection is for Pomponazzi a source of confusion as much as of confidence: man is of a nature 'not simple but multiple, not certain but ambiguous, in between mortal and immortal things'.[4] Pomponazzian 'ambiguity' is realized in the moral and spiritual complexity and confusion of Jacobean drama in England, for it is both symptom and source of a disenchantment, mirrored in pessimistic theories of the incompatibility of free will and determinism, which continued through the sixteenth century and helped to set the agenda for the seventeenth.

It has recently become clear that sophisticated treatments of the problems of determinism, freedom and responsibility cannot overlook the 1973 years of debate between the death of Aristotle and the publication of *Leviathan*. Sorabji and White in particular have demonstrated the subtlety and relevance of Hellenistic theories.[5] Meanwhile Dihle has investigated the clash between Greek conceptions of the universe as rationally ordered and Judaeo-Christian voluntarism as leading to Augustine's 'invention' of the modern notion of the will.[6] Stoic determinism, with its eternal causal chain available for man's rational examination, can be seen as the philosophical systematization of the Greek intuition noted by Dihle. It was the root of two of the three sixteenth- and seventeenth-century general classes of determinism at which I shall look in this paper; the explicitly neo-Stoic determinism of Justus Lipsius and the naturalistic determinisms of Pomponazzi and of Hobbes. These views, in contrast to the radical providentialist determinism of Luther and Calvin, failed to gain widespread acceptance. But, despite obvious major differences between them, the three determinisms were not entirely distinct, particularly in the eyes of their critics. All three were accused, as determinisms have always been, of leading to moral decay, political and religious subversion, and the erosion of human dignity. Whether denying free will like Luther, finding room for human freedom within a deterministic world of causes like Lipsius, or arguing for revised conceptions of freedom and responsibility like Pomponazzi and Hobbes, these views were attacked for not attributing a sufficient degree of flexibility to human decision and action. Determinism is never popular.

I will not always distinguish between the theological problem of freedom versus predestination and the philosophical problem of freedom versus determinism. The latter grew out of the former only gradually, and the possibility of their separation was itself a contentious issue. In addition, I do not of course intend to *identify* Luther's views with Calvin's, Lipsius' with the English neo-Stoics', or in particular Pomponazzi's with Hobbes'. The failures of these systems had at least as much to do with their manifest cultural image, and thus with their reception by generally hostile writers, as with what their proponents actually thought.

Modern work on free will often seems to assume that any determinism must be that of Laplace, tied to classical mechanics. But the intuition that the universe is one connected causal whole does not depend on any particular physical theory. Pomponazzi was not less of a determinist for believing in occult powers and systematic astrological causation: he was just working within an erroneous physical system.[7] Determinism is a blanket description used to cover many particular views. I shall mean little more by it in general than the thesis that every event has a cause and that the same cause is always followed by the same effect. A note on some other terms: compatibilism (also known as soft determinism) is the view that the truth of determinism does not rule out human freedom. Incompatibilists think that it does. Of these, hard determinists claim that determinism is in fact true, and thus that we are not free; whereas the libertarian position is that determinism is in fact false, and that we are free. This latter view was that of the most vehement philosophical opponents of sixteenth- and seventeenth-century determinism; so the paper concludes with a sketch of two libertarian attempts at positive accounts of freedom, those of Mersenne and Cudworth.

I. NEO-STOICISM AND RELIGIOUS DESPAIR

A powerful early statement of determinism is that of the Stoic Chrysippus. Fate is 'the natural order of all things established from eternity, mutually following each other in an immutable and imperishable connection'.[8] Towards the end of the sixteenth century Stoicism was reinvigorated as it became clear that eclectic use of its ethics could enhance Christianity. The oracular Justus Lipsius tried, most systematically in his *Physiologia Stoicorum* of 1604, to reconcile Stoic determinism with human freedom:[9] and an ethical neo-Stoicism became as fashionable as its ancestor had been in first century A.D. Rome.

Lipsius' God is Providence, Fate, Necessity and the Greek Logos. He notes Augustine's approval of the Stoic attribution of 'the so-called order and connection of the causes to the Will and Power of God most high'.[10] But God is no slave to Necessity; the decrees he obeys are his own.[11] With respect to human freedom, Lipsius expounds an ultra-rational compatibilism. He sees that the attempts to divorce necessitation from causation ascribed to Chrysippus[12] do not allow sufficient flexibility to human action. Chrysippus has failed in his attack on 'men who, when they have been convicted of crime and in an evil deed, flee for refuge to the necessity of Fate, as if to some kind of asylum'.[13]

But, against this, Lipsius' compatibilism is barely more than asserted.

Although God/Fate creates our character, these inbred causes can somehow be moderated or even turned aside easily ('leviter') by the Will, which is a proximate and auxiliary cause.[14] Unlike later attempts to save freedom and morality[15] Lipsius does not go so far as entirely to remove human will from the chain of universal causation: but he gives inadequate reason to assume that fully determined choice is 'free' in a sense strong enough to ground Christian ethical practice.

Lipsius' neo-Stoicism, as an attempt fully to rationalise Christian theology, was bound to conflict with the voluntarism inherent in Christianity since its inception.[16] But the ethical aspect of neo-Stoicism, in its popular form an unmetaphysical philosophy of life, found a continued popularity, among a multiplicity of unreconciled philosophical dogmas, for intellectuals who desired a Christian morality independent both of discredited Catholic authority and the faith of fanatical reformers. But almost all its adherents were Christians before they were Stoics, for as an all-embracing philosophical system its reconciliatory tactics failed to hide the inconsistencies between the two world-views. This is apparent in the problems its adherents faced in England.

Stoicism's remarkable vogue in the late Elizabethan and early Jacobean era has been well documented, particularly with regard to its literary influence.[17] Thomas James, prefacing his 1598 translation of du Vair's *La Philosophie morale des Stoiques* (1594), remarked that 'Christians may profit by the Stoicks' because 'no kind of philosophie is more profitable and neerer approaching Christianitie'.[18] Fulke Greville was one who tried to carry out this project of incorporation.[19] But his explorations of such a fusion could not but induce public criticism from those unwilling to be bound by the Stoic causal chain. One of the cynical choruses in the 1609 Quarto of Greville's play *Mustapha*, the Chorus Tartarorum, attacks 'Religion, thou vain and glorious style of weakness'. But a copy of the 1633 Folio edition in the Bibliothèque Nationale has a manuscript annotation in the hand of Sir Kenelm Digby, who would later become the first to introduce the Cartesian philosophy into England.[20] The line now reads 'Vast superstition! Glorious style of weakness', because the original 'seemed too atheistical to be licensed at the press'.[21] Internal as well as external problems beset neo-Stoic Christianity.[22] In the world of the Jacobean malcontent, where man is 'confounded in a maze of mischief, /Staggered, stark fell'd with bruising stroke of chance',[23] the idea of rational harmony with a divine and beneficently ordered scheme becomes a mirage:

> Philosophy maintains that nature's wise
> And forms no useless and unperfect thing....
> Go to, go to, thou liest, Philosophy!
> Nature forms things unperfect, useless, vain.[24]

Yet more widely criticised than Stoicism's determinist metaphysic is its ethic of patience in adversity. Even 'our English Seneca', the Anglican Bishop Joseph Hall, rejected Stoic emotionlessness: 'I would not be a Stoic, to have no passions; for that were to overthrow this inward government God hath erected in me; but a Christian, to order those I have'.[25] Webster's Antonio, in the manner of the best Stoics of Books 15 and 16 of Tacitus' *Annals*, finding easy ways to die, begs the Duchess

> O be of comfort,
> Make patience a noble fortitude:
> And think not how unkindly we are us'd.

We are not surprised to see him found unconvincing:

> Must I like to a slave-born Russian
> Account it praise to suffer tyranny?[26]

II. DIVINE FATALISM ARBITRARY

Stoic resignation was similarly rejected by the Reformation theologians. Calvin's first published work, in 1532, was a commentary on Seneca's *De Clementia*, but his praise in the preface of Seneca's 'perfect grasp of the mysteries of natural philosophy' and supremacy in ethics[27] was later displaced by an impatience with Stoic detachment. 'Ye see that patiently to bear the Cross is not be utterly stupefied and to be deprived of all feeling of pain. It is not as the Stoics of old foolishly described the "great-souled man": one who, having cast off all human qualities, was affected equally by adversity and prosperity'.[28] In particular, he was careful to distinguish his own theory of predestination from the 'fatalism' of the Stoics.[29] If God is direct cause even of every drop of rain,[30] and there is no 'wandryng power' independently inherent in any creature,[31] then both Catholic free will and rational Greek determinism fail to do justice to the phenomena. As John Knox put it, 'Fortune and adventure are the words of Paynims ... That which ye scoffingly call

Destiny and Stoical necessity ... we call God's eternal election and purpose immutable'.[32]

Besides Augustine's voluntaristic conception of God, the intellectual ancestor of Luther's denial of free will was the *De Libero Arbitrio* of Lorenzo Valla (1439), which had also asserted the requisite dependence on faith ('no-one who likes philosophy so much can be pleasing to God').[33] The Reformation assertion that the fall was ordained and that some men are made necessarily damnable seems to imply God's ultimate responsibility for the existence of evil. So unless faith is exercised, and the apparatus of praise and blame, salvation and damnation in a universe in which man is caused to sin by a wholly external force is simply accepted, the doctrine will lead to despair, doubt and atheism. At the end of the seventeenth century, Pierre Bayle explained the existence of so many bungling theological attempts to save free will: 'It is the wish to exculpate God; for it has been clearly understood that all religion is here at stake, and that, as soon as one dared to teach that God is the author of sin, one would necessarily lead men to atheism'.[34]

Luther's initial attack on the Church's doctrine of free will was answered by Erasmus in 1524, and a bitter controversy ensued, in which Melanchthon was won over by Erasmus. But Luther repeatedly pointed out Erasmus' (typically compatibilist) inconsistency in claiming that on the one hand man can do nothing without grace, but on the other the human will has enough power to fulfil its own commands and even to earn eternal life.[35] In this as elsewhere Erasmus, 'the fountainhead of the systematic deliberate vagueness of liberal Protestant theology',[36] sets the agenda for future attempts to defuse real theological controversy. Like Cudworth[37] he blames injustice and moral failings on those who teach that men are not causally responsible for their own actions. 'While we are fully occupied singing the praises of faith, we must be careful not to destroy freedom, because if we do, I cannot see how we could resolve the problems of justice and divine mercy ... who will be able to bring himself to love God with all his heart when He created hell seething with torments in order to punish His own misdeeds in His victims as though He took delight in human torments?'[38]

III. NATURALISTIC DETERMINISM: POMPONAZZI, CHRISTIANITY AND FREE WILL

Hard determinism was the rarest view in antiquity on freedom and determinism,[39] and, apart from hints in Valla, it did not gain much ground in the early Renaisisance. But in the early sixteenth century two very different new systematic philosophies denied that we have as much free will as traditional philosophy and theology had assumed. The flat message of Luther's *De Servo Arbitrio* had been that 'there can be no free will in man, in the angels or in any other creature'.[40] Because this radical determinism was tied to a strict providentialism, looking only to God as first cause, its problematizing of moral responsibility could at least be referred back to God's incomprehensible will. But in its removal of the initiative from the human will to maintain God's omniscience and omnipotence, it shared a common determinism with the system of Pietro Pomponazzi.[41] He, however, advocated no such continual meditation on God alone as first cause,[42] had, unlike the Stoics, no popular ethical system readily assimilable to Christianity, and was thus more vulnerable to violent criticism from those fearing the collapse of traditional moralities. This kind of common ground between radical determinist providentialism and radical determinist naturalism is parallel to the similar alliance noted by Keith Hutchison[43] on the issue of supernatural and natural causation. It is just as striking in the case of determinism, free will and responsibility, for the contemporaneous systems of Luther and Pomponazzi both threatened, from different directions, traditional Catholic moralities based on free will, as adopted by the Council of Trent, and as would survive into the mechanical philosophy of the seventeenth century.[44] The similarity of the two new determinist systems of the 1520s has been briefly noted by Poppi: 'In the Reformation debate, therefore, Pomponazzi's philosophical fatalism was infiltrated by a fideistic fatalism of the opposite kind'.[45]

Pomponazzi's *De Fato, De Libero Arbitrio et De Praedestinatione*, completed in 1520, is a sustained and sophisticated attack on traditional Aristotelian vagueness about the relations between causation and necessitation, or between free will and responsibility.[46] Using the *De Fato* of Alexander of Aphrodisias, an attack on Stoic fatal necessity, as his primary target, he criticises the Aristotelian exemption of chance events and coincidences from the realm of necessity.[47] Fortuitousness *is* compatible with necessity, as when a stone falls and happens to hit the head of an unwitting bystander.[48] This is exactly the kind of example Aristotle had used to avoid necessity.[49] But Pomponazzi points out that this contingency is not the real ontological indifference of the explicitly libertarian view, the actual physical possibility of an event happening or not happening; it is simply a notion we apply to things which sometimes happen and sometimes do not. (If it does rain tomorrow, it rains necessarily; if it doesn't rain, the lack of rain is necessary). 'And that is the true meaning of contingency.'[50] He neatly demonstrates the contradiction, implicit in Aristotle's works on ethics and explicit in Alexander, between the assumed libertarian possibility of self-change, which Alexander takes to be self-determination as between two incompatible choices, and the Aristotelian denial of self-change in sublunary creatures.[51] For Pomponazzi, as for Aquinas before and for Hobbes after, the will is not an independent psychological faculty. And as it cannot change itself, it must be changed by some higher, external source. Of decision, he writes 'It is held without qualification to be within the power of the will, which it in no way is'.[52]

Surprisingly, the *De Fato* was not banned by the Inquisition on its publication in the mid-sixteenth century: neither, until a century after its original composition, was Pomponazzi's *De Incantationibus*, which cast doubt on the existence of angels and demons and gave a naturalistic account of the rise and fall of religions, including Christianity.[53] But his views on moral matters arising out of fate and free will were subject to religious criticism throughout the sixteenth century.[54] His former student Paolo Giovio claimed in 1557 that Pomponazzi's doctrines led 'to the corruption of young men and the destruction of Christian discipline'.[55] But, just as in *De Immortalitate Animae* Pomponazzi had argued that the unqualified ('simpliciter') mortality of the soul does not destroy human goals and ideals,[56] so in *De Fato* he makes a case for a moral responsibility which could be compatible with universal necessitation. We are Fate's children, and ordinary praise and blame are out of place.[57] But a revised conception of morality can see good and evil both as parts of the natural order. Indeed, as for the Gnostics, good *requires* evil: 'It is necessary that there should be sin: providence intends there to be sin and is itself author of sins'.[58] He avoids the ensuing temptation to blame God for evil by claiming that God's behaviour towards man is as free and blameless as is man's towards cattle and chickens.[59] Given the natural existence of evil, Pomponazzi suggests that judgements about good and evil can function as do judgements on good wine or noxious insects.[60]

Book Three of the *De Fato* begins a different and incompatible account of freedom which is firmly Christian, taking free will as a premise.[61] As Copenhaver notes, however, 'in the larger context of his work these attempts to repair the damage done to free will ring hollow'.⁶² Pomponazzi has already counterattacked against Christian doctrines of free will in Book Two by praising Stoic determinism for refusing to deny any powers to God. The Stoics 'preferred to be servants and followers than to be impious and blasphemers; they believed that everything was fated and arranged by providence and that there is nothing in us which is not done by providence'.⁶³ So Pomponazzi too professes docility, using the same phrase.⁶⁴ This attribution of all causes directly to God is not quite an accurate description either of Stoic determinism or of Pomponazzi's version of it.⁶⁵ But he had to tread carefully. Less than a hundred years later, in February 1619, Vanini was burnt in Toulouse for expounding Pomponazzian naturalism. Poppi's remark that the Christian account of free will in Book Three of the *De Fato* was a 'dialectical line of defence'⁶⁶ is confirmed by a close reading of the wonderfully ironic epilogue to the whole work.⁶⁷ It ends with Pomponazzi's acknowledgement that because the Church has condemned Stoic fatalism, he too must deny it, 'and the Church is firmly to be believed'.⁶⁸ Against anticipated attacks on his work, Pomponazzi happily issues a disclaimer of its doctrines; 'Moreover, of the opinions I have put forward, I adhere only to as many as the Roman Church, to which both in this and in other matters I wholly submit, will have approved'.⁶⁹ But a few paragraphs earlier, at the beginning of the epilogue, Pomponazzi had concluded that, although no account of fate and free will is wholly satisfactory, that of the Stoics, in nature alone and by reason, is 'furthest removed from contradiction'.⁷⁰ Here he says that the best argument against it is that it makes God the cause of sin. Pomponazzi remarks that this consequence 'seems ['*videtur*'] absurd and erroneous',⁷¹ before referring back to his own arguments in Book Two which remove its sting. He provides just sufficient disclaimers throughout to escape more than unofficial censure. A clever man.

IV. NATURALISTIC DETERMINISM: HOBBES, MATERIALISM AND MORALITY

Naturalism, materialism and determinism are thought to be entirely compatible by many modern philosophers, aspects of the same broadly 'naturalistic' perspective. Cudworth, unlike Bramhall, saw the crucial link between determinism and materialism in Hobbes' thought. In the now fragmentary *Discourse of Liberty and Necessity* Cudworth claims that Hobbes denied free will because he 'denied all spirituality and immateriality and made all cogitation, intellection and volition be nothing but mechanical motion and passion from objects without ... wherefore it is a sufficient confutation of (Hobbes) to show that there is another substance in the world besides body',[72] which Cudworth, of course, thinks he has done in *The True Intellectual System of the Universe*.[73]

Hobbes' deterministic psychology has been the focus of much work generically classifiable as 'Hobbism'[74] both in the seventeenth century and again recently, as materialist theories of mind have gained widespread acceptance. Cognition is reduced, via sensation, to motion. Deliberation is a vector of mechanical motions, of which the will is the resultant vector. Mental processes are 'nothing really, but motion in some internal substance of the head',[75] and psychology is reducible to the mechanics of appetite and aversion.[76] This Hobbes is at least a revisionary if not an eliminative materialist.[77] Psychological Hobbism has certainly over-emphasised the *external* determination of mental events: Jamie Kassler in this volume demonstrates the importance of the movement outward in Hobbes' physiology, and thus of the internal determinism of action and character. It would indeed be a crude materialist determinism which ignored (deterministic) inner processes. But what is at issue here is whether the naturalism in Hobbes brought out by Kassler is at odds with the materialism of contemporary and modern Hobbist interpretations of Hobbes. I tend to think not;[78] but the following treatment of Hobbes' determinism holds for both readings.

Hobbes knew Pomponazzi's thought: his friend Mersenne had devoted the first section of his *Quaestiones Celeberrimae in Genesim* to an attack on Pomponazzi's and Vanini's naturalistic denial of the immortality of the soul and the existence of angels and miracles.[79] There are interesting similarities between the two determinisms. Hobbes' account of the will as merely the last desire or appetite in deliberation is in accord with Pomponazzi's unification of intellectual and sensitive soul. Both men offer an argument against free will from human ignorance of causes: in our epistemically deprived state we do not know the true causes of things, and we pass easily from such ignorance to the illusory belief that there are no such causes in nature. This account, which may go back to the ancient atomists,[80] is intended to explain *away* our intuitions of free will.

Pomponazzi's account of soul and mind is very complex, trying to ward off the crudest varieties of materialism; but at the very least he too claimed that the human soul was by nature absolutely ('simpliciter') 'materiale' and only relatively ('secundum quid') 'immateriale'.[81] Of course not all causes are *easily* explicable in physical terms; Pomponazzi accepted astrological causation, and explained alleged phenomena in terms of occult or hidden powers and qualities. This could be seen as an area in which Hobbes is typical of a general advance on Pomponazzi and other Renaissance naturalists and magicians. But Pomponazzi's acceptance of occult qualities is not naturalism in an antimaterialist sense.[82] The shortcomings of Aristotelian physics and theories of matter make such an inability to tolerate temporary ignorance unremarkable in the Aristotelian philosopher desiring to know the causes of things. Hobbes, in contrast, shares with many philosophers of the scientific revolution an acceptance that all explanation is incomplete. Keith Thomas pinpoints this as an important intellectual and cultural innovation in the seventeenth century.[83] But despite this change, it is unnecessary to see belief in occult qualities as marking an 'irreconcilable difference' between Renaissance naturalism and seventeenth century philosophy, as Keith Hutchison has shown.[84] Seventeenth century science did not so much reject occult qualities as break down the distinction between occult and manifest, by showing occult qualities to be no more and no less intelligible than any other causes, and subsuming those which gained scientific respectability into the new science. Hutchison's remark on natural magic and the new science in general could be applied to Pomponazzi and Hobbes in particular: '. . . the two systems have in common a willingness to deal with occult qualities and a refusal to accept that insensibility implies spirituality: it is within natural magic that we can find precedents for the confidence with which seventeenth century philosophy insisted that the insensible realms of nature could be profitably entered by human thought.'[85]

I noted earlier Poppi's drawing of the parallel between the two new determinisms, Pomponazzian and Lutheran, of the 1520's. He follows this with the comment that both are 'equally destructive of man's reality; in the first case because he is the victim of material cosmic forces, and in the second because he is the victim of a predestined will. Man's highest faculties are systematically demoted and denied; his works are entirely disregarded, and his moral commitments discarded as illusory'.[86] Similar remarks are easy to find in both seventeenth and twentieth century treatments of Hobbes' determinism.[87] But few if any determinists, least of all Hobbes, have ever been amorally Hobbist in the sense their opponents assume they must be. Alternative determinist conceptions of self-creation, and revised freedoms and moralities, are consistently ignored. We need historical and cultural explanations of the intuitions behind the reception in England of Hobbes' 'blasphemous, desperate, and destructive opinion of fatal necessity'.[88] One of the ironies of Hobbes studies is that Hobbes 'argued in support of a social and political order the conceptual resources to justify which he had removed'.[89] But free will, unlike some of the other concepts mentioned by Shapin and Schaffer in this context, was not so much a legacy of an existing social and political order which was about to disappear, as an important factor in the construction of an emergent new order which required the creation of an idealized autonomous individual subject.[90] On Hobbes in particular, Mintz has described Bramhall's and others' criticisms of the 'ethical inconveniences' of determinism.[91] Benjamin Laney complained in 1677 that it 'must needs shake not only the Foundation of all Religion, but even of humane Society'.[92] The prevalence of these assumptions about the consequences of determinism needs to be related to an English fear of social and religious corruption among intellectuals which was not confined to political reactionaries:

> If this fail,
> The pillared firmament is rottenness,
> And earth's base built on stubble.[93]

Determinism and naturalism are a short step from atheism. As Sir William Alexander wrote in 1630, 'Young Naturalists oft old Atheists doe prove'.[94] Bramhall's epistle to the reader, in his *Vindication of True Liberty from Antecedent Extrinsical Necessity*, gives us further suggestions on the social implications of determinism: Hobbes' 'principles are pernicious both to piety and policy, and destructive to all relations of mankind, between prince and subject, father and child, master and servant, husband and wife; and they who maintain them obstinately, are fitter to live in hollow trees among wild beasts, than in any Christian or political society. So God bless me'.[95]

Hobbes' particular defences of moral practices in a deterministic world are generally consequentialist.[96] There are extrinsic justifications for praise and blame. Even 'retributive' punishment does not require full moral responsibility, for it works 'to the end that the will of men may thereby be the better disposed to obedience'.[97] Rejection of Stoic rational design *and* Reformation providentialism had always been recognised as leading to ethical relativism:

> Most things that morally adhere to souls
> Wholly exist in drunk opinion,
> Whose reeling censure, if I value not,
> It values nought.[98]

Hobbes accepts this, but claims that things can nevertheless be necessary and yet praiseworthy, just as they can be necessary and yet dispraised.[99] Consequentialism in some form is surely the best ethical framework for a sincere determinist, and if there is a way out of the determinist maze described in this paper this is it. But it makes moral *theorizing* desperately difficult (rightly so?), and no libertarian has ever been convinced that it could be a genuine substitute for moral realism. One reason for this is that it fails to save, to any significant degree, what the libertarian considers to be the moral phenomena. Given this, it is slightly odd that Hobbes is supposed to be a compatibilist.[100] He acknowledges that dispute over questions of free will and determinism among 'the greatest part of mankind, not as they should be, but as they are ... will rather hurt than help their piety', that he would not be putting forward his argument if Bramhall had not provoked him, and that he hopes 'your Lordship and his will keep it private'.[101] Free will should still be defended in public. Bacon had written to some judges in 1617 that 'there will be a continual defection, except you keep men in by preaching, as well as the law doth by punishing'.[102] Despite regular protestations of innocence, Hobbes knew the revisionary consequences of his determinism for morality as well as did his critics: but, unlike them, he was willing to embrace what they saw as ethical inconveniences.

V. JOHN WEBSTER AND THE FAILURES OF DETERMINISM

Hobbes' account of contingency as ignorance of causes is justifiably famous:

A wooden top that is lashed by the boys, and runs about sometimes to one wall, sometimes to another, sometimes spinning, sometimes hitting men on the shins, if it were sensible of its own motion, would think it proceeded from its own will, unless it felt what lashed it. And is a man any wiser, when he runs to one place for a benefice, to another for a bargain, and troubles the world with writing errors and requiring answers, because he thinks he doth it without any cause other than his own will, and seeth not what are the lashings that cause he will?[103]

The image, as used in this context, is not his own. We have already seen how the Duchess' rejection of Antonio's advice to 'make patience a noble fortitude' is symptomatic of embarrassment over Stoic determinism and impatience with Stoic resignation. She continues with an appeal to a radical providentialism, to God as guide of *all* human affairs:

> And yet, O Heaven, thy heavy hand is in it.
> I have oft seen my little boy scourge his top,
> And compared myself to't: nought made me e'er go right,
> But Heaven's scourge-stick.[104]

This is one of Webster's rare borrowings from Sidney's verse works. Reference to this source for the image makes its providentialist inclination clearer:

> Griefe onely makes his wretched state to see
> (Even like a toppe which nought but whipping moves)
> This man, this talking beast, this walking tree ...
> But still our dazeled eyes their way do misse,
> While that we do at his sweete scourge repine,
> The kindly way to beat us to our blisse.[105]

Bramhall too, in perhaps his most famous rhetorical flourish against Hobbes' determinism, asserting that man *must* be more than talking beast or walking tree, echoes a Webster borrowing from Sidney. Bosola notoriously complains, after the unintended death of Antonio, that 'We are merely the stars' tennis balls, struck and banded/Which way please them'.[106] Similarly Hobbes' doctrine 'destroys liberty, and dishonours the nature of man. It makes the second causes and outward objects to be the rackets, and men to be but the tennis-balls of destiny'.[107] A whole conflation of traditions is at work here. The idea itself is an old conceit, going back at least to Plautus.[108] It reaches Webster through a use by Sidney which mingles the medieval morality tradition with a Calvinist belittling of man's powers: 'in such a shadowe, or rather pit of darkness, the wormish mankinde lives, that neither they know how to foresee, nor what to feare: and are but like tenisballs, tossed by the racket of hyer powers'.[109] The Jacobean malcontent is the bruised inheritor of what Burton called this 'horrible consideration of the diversity of religions which are and have been in the world'.[110]

In these texts are inscribed the failures of the three major determinist systems of the early modern age. The Duchess rejects Antonio's Stoicism; her own frail providentialist faith in God's scourge-stick proves unfounded; Hobbes' use of the image of the top to explain away our intuitions of free will is rejected because it gives man no more than brute or object status.[111] What accounts of human freedom could be offered in their place? In England the Restoration government banned public preaching and discussion on the topic of free will, among other subversive issues,[112] as the Royal Society took steps towards the correcting of excesses in natural philosophy, without Hobbes.[113] But besides censorship, the seventeenth century saw two criticisms of determinism from major new philosophical directions, the mechanical philosophy and Cambridge Platonism. We will take Mersenne's criticism of Pomponazzian naturalism and Cudworth's deft but desperate bolstering of free will against the evils of Hobbist atheism as representative.

VI. LIBERTARIANS ON FREE WILL: MERSENNE

The religious dangers of overzealous application of the mechanical philosophy were obvious even before Hobbes' perniciously materialist version. So sometimes, 'to preserve religion, morality and science',[114] it was 'more prudent to adopt the mechanical philosophy in an attenuated form even at the cost of philosophical untidiness or inconsistency'.[115] Mersenne died before the publication of the Hobbes-Bramhall debate. But his attacks in *Quaestiones Celeberrimae in Genesim* and *L'Impiété des Déistes*[116] on the naturalisms of Pomponazzi, Cardano and Vanini show that, like Bramhall, he was willing to exclude the human mind and in particular the will from the mechanical universe of secondary causes. The difficulty of Mersenne's limited defence of supernaturalism against Neoplatonic magic and astrology on the one hand and the naturalists' denial of angels and miracles on the other has been demonstrated by Hine: 'with naturalism, Mersenne's task was to explain the limitations of nature. With magic, he had to emphasise the limits of supernatural events and angelic powers'.[117] A similar balancing act is apparent in his attitudes to determinism and free will. He criticizes Pico and Ficino for attributing too much to human freedom.[118] But the threat of naturalism, in contrast, is the erosion of human dignity by cosmic destiny. Mersenne's typically libertarian assumption is that the determinists' denial of absolute self-determination automatically removes all justification for any moral striving whatsoever. A naturalistic Averroist view of religions as natural phenomena like any other, as taken up by Pomponazzi,[119] ensures that 'whether one preserves the name of "free will" or not, man does not escape his destiny'.[120]

There is little logical space for Mersenne's position. He attacks the Cabala and astrology because they fail to do the explanatory and predictive work they profess to,[121] and because they derive from disreputable Eastern sources. But his alternative, 'Greek' conception of the cosmos as rationally ordered, and the importance for mechanism in general of the idea of a nature subject to intelligible laws, seem prima facie to suggest a naturalistic determinism which denies the belief in miracles and the efficacy of prayer to which the good Catholic priest is committed. When these opposing influences, religious morality and mechanistic science, come into conflict, the outcome is decided in a familiar manner, by the necessity of making man rather than God responsible for sin. Despite, or perhaps because of, mechanism's boast to be the true science to deliver politic society from the variety of false sciences, Mersenne must in the end profess the traditional Catholic liberty of indifference: 'So the will, then, is in my opinion able to pursue either one of two objects equally set before it, even if no greater reason should be apparent to it why it should pursue one rather than the other'.[122] He supported free will against the Jansenists in the early 1640s.[123] Even Lenoble confesses Mersenne's difficulties in attempting 'the impossible synthesis ... of two violently opposed traditions; the ancient tradition which identifies God and destiny, and which, with regard to man, must thus subordinate freedom to nature; the Christian tradition for which God gives benevolently and freely, and creates as his masterpiece souls which are truly free'.[124] These are *exactly* the two incompatible tendencies identified by Dihle as contributing to the origins of the modern problems of the will.[125]

This is another angle on the problems created by the mechanists' tendency to remove the mind from the realm of physical causality. If 'voluntary' human action is not to be explained within the same causal nexus as the behaviour of physical bodies, how does the libertarian freedom of indifference possessed by a separate faculty of the will itself help in giving an account of the springs of action which preserves a space for rational deliberation? The arbitrariness which dogs any account of action, from Lucretius to J. R. Lucas, which does not refer to previous deterministic causal factors, infects Mersenne as well. In modern terms, 'if no amount or kind of cognitive and volitional capacity and complexity that could obtain in a deterministic world will suffice for free agency, then adding a requirement of indeterminism won't help'.[126]

VII. LIBERTARIANS ON FREE WILL: ANGLICANISM AND CUDWORTH

The elements of Reformation determinism which were fully absorbed into the Anglican mainstream were nevertheless somewhat less menacing than the varieties developed in Geneva, Scotland or New England: hell was sanitized, and, escaping the 'bruising stroke of chance', the English found God's scourge-stick the *kindly* way to beat them to their bliss. Webster has Antonio attempt to bolster the Duchess' quavering faith in providence.

> Do not weep:
> Heaven fashioned us of nothing; and we strive
> To bring ourselves to nothing.[127]

The tag derives from Donne's *First Anniversarie*

> Wee seeme ambitious, God's whole worke t'undoe;
> Of nothing hee made us, and we strive too,
> To bring ourselves to nothing backe.[128]

Self-denigration was the darker side of Anglicanism, explicit at least before Laud made belief in free will an obligatory article of faith. It mixed with a more optimistic if less honest English compatibilism and complacency. The general attitude is apparent in Hooker's complaint: 'A number there are, who think they cannot admire as they ought the power and authority of the word of God, if in things divine they should attribute any force to reason. For which cause they never use reason so willingly as to discredit reason'.[129]

The first years of the seventeenth century saw a climax of religious pessimism, apparent in the raw nerves touched by the Jacobean dramatists. In Holland, the controversy over Arminian attacks on Calvinist predestination was brought to a temporary halt by the death of Arminius himself in 1609 and by the victory in 1608–9 in front of the Dutch States-General of the followers of Gomarus (who had studied at Oxford and Cambridge). These 'supralapsarians' reiterated the Calvinist claim that God wills the fate of each man before creation.[130] This was the dominant view in England too, indeed almost universal before the accession of Charles I.[131]

But alongside predestinationism in England were prevalent optimistic views of causation and free will. Peter Baro, Professor of Divinity at Cambridge from 1574 to 1596, believed that 'God willed that there should be divers and sundry causes, namely some necessary and othersome also free and contingent: which according to their several natures might work freely and contingently, or not work. Whereupon we conclude that secondary causes are not enforced by God's purpose and decree, but carried willingly and after their own nature'.[132] The tendency towards a libertarian view of the autonomy of the human will among other secondary causes evident in this passage coexisted happily with a genuinely prescientific devaluation of secondary natural causes; 'thus must we in all things that be done, whether they be good or evil (except sin, which God hates and causes not), not only look at the second causes, which be but God's means and instruments whereby he works, but have a further eye and look up to God'.[133] For the Elizabethan preacher, distanced from theological and scientific controversy on the continent, second causes were of little interest anyway. But this latter view does show the influence of Calvinist determinism: Calvin too could allow talk of secondary causes, but affirmed that 'whatsoever instruments God uses, his original workyng is nothing hindered thereby', for 'we set no power in creatures. Onely this we say, that God useth meanes and instruments whiche he hymselfe seeth to be expedient'.[134]

In England, religious despair arising from radical providentialism and from the failure of Stoicism was often an impetus to construct a less threatening metaphysic. Some, like Donne and his Oxford near-contemporary Marston, found solace in a retreat to the Anglican altar. Ignoring or suppressing the problem, like the Restoration censors, was one way out (Webster's Duchess, in our text, knows the perils of thinking: 'All our wit/And reading brings us to a truer sense/Of sorrow'[135]). But not the only way. One of the most resolute supralapsarian opponents of free will in the late sixteenth century was the Cambridge preacher William Perkins. His chosen successor as vicar of St. Andrews'

Church in Cambridge was his collaborator in divinity, one Rafe Cudworth. We know that his son Ralph's first reading of the ancients initiated crisis and revolt against the strict Calvinism of his upbringing.[136] If men accept that they have no free will and that God is cause of all, debauchery, scepticism and infidelity, thinks the natural libertarian, can be the only results: Nashe in 1592 had his character Ver in *Summer's Last Will and Testament* complain, after surveying the world's evils, 'If then the best husband be so liberal of his best handiwork, to what end should we make much of a glittering excrement, or doubt to spend at a banquet as many pounds as He spends men at a battle?'[137] Ralph Cudworth includes the Reformation 'Theologick Fate' and the ideas of that 'atheistic politician' whom he never mentions by name, Hobbes, among his four atheistical doctrines of the Fatal Necessity of all Actions and Events.[138] Just as the Calvinist Divine Fatalism Arbitrary makes God 'meer arbitrary will omnipotent',[139] so a denial of free will on the basis of an overenthusiastic mechanism, looking only to second causes, fails to explain the alleged (moral) phenomena, and leaves us with no 'measure or norma in nature'.[140] So instead Cudworth, with a phrase reminiscent of the chameleon-like Renaissance man of Pico della Mirandola, upholds man's 'potential omniformity'.[141]

But Cudworth, unusually, is not content with the taking of free will as a given, and just asserting *that* it is. Mintz has unravelled the tortuous positive account of *what* it is in the manuscripts of Cudworth's *Discourse of Liberty and Necessity*.[142] Agreeing with Hobbes against Bramhall that the will is not a separate faculty, he sees the free will as the soul redoubled on itself, as giving the soul *sui potestas* over itself and the ability to 'command it Selfe or turne it Selfe this way and yt way'.[143] But although he professes not to be avowing an arbitrary freedom of indifference,[144] he found no easy road between that and determinism. The perennial objection to compatibilism that choice must be either determined by external and internal causes beyond the individual's control, or be arbitrary, forced him finally into incompatibilism, accepting 'indifferent Voluntaneity' almost despite himself as the root of sin.[145] Because he will accept only an 'eternal and immutable' morality,[146] he must finally retreat to an incoherent libertarianism. If determinism is true, the reasons why particular determinisms have failed might be exactly those intuitions about freedom which have to be changed if they cannot be satisfied. But, as the case of Cudworth shows, nothing but a significantly revised ethics has any prospect of survival in a deterministic world.

NOTES

* Many thanks to Jamie Kassler and Stephen Gaukroger for comments on an earlier draft of this paper.
[1] Webster, J., *The White Devil* (1612), V.vi.177—8. References to Webster are to the Revels editions, edited by J. Russell Brown, *The White Devil*, London (1960) and *The Duchess of Malfi*, London (1964).
[2] Webster, *The Duchess of Malfi* (1614), I.i.359—61.
[3] Milton, J., *Paradise Lost*, London (1667), II.557—61.
[4] Pomponazzi, P., *De Immortalitate Animae*, Leyden (1534), p. 5.
[5] Sorabji, R., *Necessity, Cause and Blame*, London (1980); White, M., *Agency and Integrality*, Dordrecht (1985).
[6] Dihle, A., *The Theory of Will in Classical Antiquity*, Los Angeles and London (1982).
[7] Similarly, if a deterministic theory should be found to underlie quantum theory at a more fundamental level of physical reality, this of course would not entail a return to the world-view of classical physics.
[8] Chrysippus at Aulus Gellius, *Noctes Atticae* vii.2, quoted by Lipsius, J., *Physiologia Stoicorum*, in *Opera Omnia*, 4 Vols, Wesel (1675), IV. 860.
[9] *Op. cit.* (note 8). See Saunders, J., *Justus Lipsius: The Philosophy of Renaissance Stoicism*, New York (1955). On attitudes to Stoicism in the earlier Renaissance see Schmitt, C., and Skinner, Q., (eds.), *The Cambridge History of Renaissance Philosophy*, Cambridge (1988), especially Kraye, J., 360—370, and Poppi, A., pp. 641—650.
[10] Augustine, *City of God*, v. 8, quoted by Lipsius, *op. cit.* (note 8), IV. 862.
[11] Much of Lipsius' effort is spent in warding off the problem of evil. He attacks Chrysippus for failing to absolve God from evil (*op. cit.*, note 8 IV. 866). Lipsius' solution is to add to his Stoicism the Gnostic/Neoplatonic account of matter as intrinsically evil (IV. 862—73).
[12] Cicero, *De Fato* 39, 41; see Sorabji, *op. cit.* (note 5), chapters 3 and 4.
[13] Chrysippus at Aulus Gellius, *Noctes Atticae* vii.2, quoted by Lipsius, *op. cit.* (note 8), IV. 866.
[14] *Ibid.*; and see Saunders, *op. cit.* (note 9), pp. 143, 147—51.
[15] See below, sections VI and VII.
[16] This point is well made by Long, A., 'Hellenistic Philosophy and the Classical Tradition', conclusion to *Hellenistic Philosophy*, London (1974), pp. 232—48; on Lipsius, pp. 239—40.
[17] Monsarrat, G., *Light from the Porch: Stoicism and English Renaissance Literature*, Paris (1984), and, on 'Lipsius his hopping style', Williamson, G., *The Senecan Amble*, London (1951), pp. 121—49. Salmon, J., 'Stoicism and Roman Example: Seneca and Tacitus in Jacobean England' *Journal of the History of Ideas* **50**, 199—225, (1989) makes an important case for the seditious political implications of a taste for Stoic literary and historical themes, particularly among the Essex circle.

[18] Du Vair, G., *The Moral Philosophie of the Stoicks*, trans. T. James, edited by R. Kirk, New Brunswick (1951), p. 45.
[19] Rebholz, R., *The Life of Fulke Greville*, Oxford (1971), especially pp. 24—6, 82—5, 219.
[20] Digby, K., *Two Treatises*, London (1644), repr. New York (1978).
[21] Greville, F., *Mustapha*, in *Selected Writings*, edited by J. Rees, London (1973); Hilton Kelliher, W., 'The Warwick Manuscripts of Fulke Greville', *British Museum Quarterly* 34, 107—21 (1970); Dollimore, J., *Radical Tragedy*, Brighton (1984), pp. 128, 281—2.
[22] On Greville's own concern with the problem of evil, see Rebholz, *op. cit.* (note 19), pp. 25, 308—9.
[23] Marston, J., *Antonio's Revenge* (1600), ed. G. Hunter, London (1966), IV.i.56—7.
[24] Marston, J., *Antonio and Mellida* (1599), ed. G. Hunter, London (1965), III.i.27—8, 34—5. At *Antonio's Revenge* II.ii.47ff. Antonio actually reads from Seneca's *De Providentia*, only to reject it as 'naught/But foamy bubbling of a fleamy brain'; cf. *The Malcontent* (1603), ed. G. Hunter, London (1975), III.i.20—8. Aggeler, G., in 'Stoicism and Revenge in Marston', *English Studies* 51, 507—17, (1970) argues that *The Malcontent* is consistently and successfully both Christian and neo-Stoic, the tragicomic mode reflecting Marston's belief that this new fusion could not accommodate blood vengeance. This seems to me to ignore the terrible ambiguities in Marston's treatment of the views put forward by the protagonist Malevole/Altofront. It is far from clear that the play upholds belief in a rationally and beneficently ordered nature.
[25] Hall, J., *Works*, 10 Vols, ed. P. Wynter, Oxford (1863), VII. 457. Sams, H., in 'Anti-Stoicism in Seventeenth- and Early Eighteenth-century England', *Studies in Philology* 41, 65—78 (1944) saw the Stoics' suppression of passion, their paganism, and their over-reliance on reason as the three main elements of the popular received perceptions of neo-Stoic views. His opinion (77—8) that the first of the three was the most prevalent and specific source of criticism in England is supported by literary sources as well as the philosophical and theological writers he cites. (I owe this reference to Jamie Kassler). This received view of Stoicism was a far remove from Lipsius' systematic philosophy let alone from actual ancient Stoic tenets. One reason for this may be that, as Salmon, *op. cit.* (note 17), 224, notes, the rational blending of metaphysics and ethics with politics which was possible, at least in theory, for Lipsius, was never a practical option for 'those English malcontents who devised their own blend of Senecan and Tacitean influence under the pressure of plots, rivalries and disappointments in the first decade of the seventeenth century'.
[26] Webster, J., *The Duchess of Malfi*, III.v. 72—4, 76—7.
[27] Battles, F. L., and Hugo, A., *Calvin's Commentary on Seneca's De Clementia*, Leiden (1969).
[28] Calvin, J., *Institutes*, trans. Beveridge, H., 2 Vols, London (1949), III.viii.9.
[29] *Ibid.*, I.xvi.8.
[30] *Ibid.*, IV.xiv.17.
[31] *Ibid.*, I.xvi.7.
[32] *The Works of John Knox*, ed. D. Laing, Edinburgh (1846—64), V. 32, 119.
[33] Valla, L., *De Libero Arbitrio*, ed. M. Anfossi, Florence (1934); quoted by Poppi, *op. cit.* (note 9), p. 650.

[34] Bayle, P., *Dictionnaire historique et critique*, Paris (1697), art. Pauliciens, rem. (I), p. 632.
[35] Poppi, *op. cit.* (note 9), p. 664.
[36] Walker, D., *The Decline of Hell*, Chicago and London (1964), p. 193.
[37] See below, Section VII.
[38] Erasmus, D., *De Libero Arbitrio Diatribe sive Collatio* in *Opera Omnia*, 10 Vols., ed. J. Leclerc, Leiden (1703—6), repr. Hildesheim, (1961—2).
[39] Sorabji, *op. cit.* (note 5), p. 87.
[40] Luther, M., *De Servo Arbitrio*, in *Werke*, Weimar (1883), XVIII, p. 786.
[41] Pomponazzi, P., *De Fato, De Libero Arbitrio et De Praedestinatione*, ed. R. Lemay, Lugano (1957).
[42] At least, not in the same part of his work. See below.
[43] Hutchison, K., 'Supernaturalism and the Mechanical Philosophy', *History of Science* **21**, 297—333, e.g. 311, (1983).
[44] See below on Mersenne, Section VI.
[45] Poppi, *op. cit.* (note 9), 666.
[46] Pomponazzi, *op. cit.* (note 41). See Huby, P., 'The First Discovery of the Freewill Problem', *Philosophy* **42**, 353—62, (1967) on Aristotle's failure to address the issues.
[47] *Alexander of Aphrodisias on Fate*, ed. R. Sharples, London (1983).
[48] Pomponazzi, *op. cit.* (note 4), I.6.
[49] Aristotle, *Physics* II.5, *Metaphysics* VI.3. See Sorabji, *op. cit.* (note 5), chapter 1.
[50] Pomponazzi, *op. cit.* (note 41), I.7.2.23 (p. 40).
[51] *Ibid.*, II.5.59 (p. 183).
[52] *Ibid.*, I.9.3.3. (p. 160).
[53] *De Naturalium Effectuum Causis sive de Incantationibus*, Basel (1567), especially chapter 13: *De Immortalitate Animae*, *op. cit.* (note 4), p. 123. Del Rio, M., *Disquisitionum Magicarum Libri Sex*, Mainz (1624), p. 10.
[54] Poppi, *op. cit.* (note 9), p. 660.
[55] Giovio, P., *Elogia Doctorum Virorum*, Antwerp (1557), p. 154.
[56] *Op. cit.* (note 4), chapters 13 and 14.
[57] *Op. cit.* (note 41), I.11.3. 35—46 (pp. 78—81).
[58] *Ibid.*, II.7.1.34 (p. 203).
[59] *Ibid.*, II.7.1.42 (p. 205); cf. II.7.1.39 (p. 204). This comparison functions as a reductio ad absurdum of determinism for the libertarian defender of free will and divine and human dignity: see Bramhall's remarks on brutish liberty in *The English Works of Thomas Hobbes of Malmsbury*, ed. W. Molesworth, London (1839—45), V. 40; and below, section IV.
[60] *Op. cit.* (note 41), I.11—15. Compare a modern aesthetic justification of 'moral' judgements in a world of hard determinism: 'I could regret being selfish or dishonest in the way I regret having no talent for music or for sport. I could judge my actions aesthetically as admirable or appalling; and these thoughts could be charged with feelings'. Glover, J., 'Self-Creation', *Proceedings of the British Academy* **69**, 445—71 (1983); cf. *I: The Philosophy and Psychology of Personal Identity*, London (1988), chapter 18.
[61] At III.6.8 (p. 252), Pomponazzi even sees the will as a prime mover.
[62] Copenhaver, B., in Schmitt, C. and Skinner, Q., eds., *op. cit.* (note 9), p. 273.

⁶³ Pomponazzi, *op. cit.* (note 41), II.1.76—7 (p. 154).
⁶⁴ *Ibid.*, II.3.19 (p. 163).
⁶⁵ The attempt to claim for naturalistic Stoic determinism the prime doctrinal advantage of radical providentialism is a sleight-of-hand in that the Stoic God is identical with (even if not bound by) at least a logical Necessity which the Reformation theologians saw as itself subject to God's Will. However, Pomponazzi's fundamental point, that the greater the power and control attributed to the human will, the more God's omnipotence is eroded, was shared by Luther and Calvin.
⁶⁶ Poppi, *op. cit.* (note 9), p. 659.
⁶⁷ Pomponazzi, *op. cit.* (note 41), Epilogus sive Peroratio, (pp. 451—4).
⁶⁸ *Ibid.*, p. 453.
⁶⁹ *Ibid.*, p. 454.
⁷⁰ *Ibid.*, p. 451.
⁷¹ *Ibid.*, p. 451.
⁷² British Museum Additional Manuscripts 4979, fol. 61: see Mintz, S., *The Hunting of Leviathan*, Cambridge (1962), pp. 127—8.
⁷³ Cudworth, R., *The True Intellectual System of the Universe*, 2 Vols, London (1678), repr. New York (1978).
⁷⁴ A good account of what constitutes 'Hobbism' (although his emphasis is on politics) is Lamprecht, S., 'Hobbes and Hobbism', *American Political Science Review* **34**, 31—53, (1940).
⁷⁵ *Human Nature*, in *English Works, op. cit.* (note 59), IV. 31.
⁷⁶ Spragens, T., *The Politics of Motion*, London (1973), pp. 68—73; Lott, T., 'Hobbes' Mechanistic Psychology', in J. G. van der Bend, (ed.), *Thomas Hobbes: His View of Man*, Amsterdam (1982), pp. 63—75. The mechanistic materialist reading of Hobbes in the twentieth century derives from Brandt., F., *Thomas Hobbes' Mechanical Conception of Nature*, Copenhagen and London (1928).
⁷⁷ For the terms, see Churchland, P. S., 'Replies to Comments'. *Inquiry* **29**, 247—8, (1986). For the modern theory, see for example, Churchland, P. M., *Matter and Consciousness*, 2nd edition, Cambridge, Mass. (1988), p. 43—9. The conceptual affinity between eliminative materialism and hard determinism is noted by Armstrong, D., 'Recent Work on the Relation of Mind and Brain', in G. Floistad, (ed.), *Contemporary Philosophy, Vol. 4*, The Hague (1983), pp. 53—4.
⁷⁸ The important passages in the later work of Hobbes are *Body* chapter XXV, in *English Works, op. cit.* (note 59), I.287—410; *Human Nature* chapters II and VII, in *English Works*, IV.3—9, pp. 31—4; *Leviathan* chapters 1, 2, 3 and 6, ed. C. Macpherson, Harmondsworth (1968), pp. 83—99, 118—30. Nothing in these passages seems to me incompatible with a sophisticated materialism: for instance, in *Leviathan* chapter 3, an initially crude associationism about 'mental discourse' is supplemented by an account of the influence of 'desire and design' on the consequence or train of thoughts. Of course any materialist will avoid fatalism by allowing that processes inside us, which form a part of a chain of physical causes, are among those which determine future actions and events. To what extent they are 'in our power', and how the self acts through a rational will when neither self nor will has any separate existence apart from the same physical causal nexus, are further questions. In psychology at least, perhaps, there is no Hobbist smoke without Hobbes' own fire.

[79] Mersenne, M., *Quaestiones Celeberrimae in Genesim*, Paris (1623).
[80] Aristotle, *Physics* 196 a1—3, b6—7 on the atomists' views on chance.
[81] Pomponazzi, *De Immortalitate Animae, op. cit.* (note 4), pp. 75—6.
[82] It is perhaps, rather, comparable to the views of those physicists of strong determinist inclination (including Einstein, Schrödinger and Planck) who have believed that indeterministic quantum theory is incomplete, or that there may be hidden variables yet to be discovered in apparently indeterministic systems. See Bohm, D., *Causality and Chance in Modern Physics*, 2nd edition, London (1984), and *Wholeness and the Implicate Order*, London (1980). Occult qualities, like hidden variables, were postulated to explain phenomena which contemporary natural philosophy could not, while avoiding recourse to supernaturalism or non-material entities.
[83] Thomas, K., *Religion and the Decline of Magic*, Harmondsworth (1973), pp. 785—94. He quotes Sir Robert Filmer: 'There be daily many things found out and daily more may be which our forefathers never knew to be possible', *An Advertisement to the Jury-Men of England*, London (1653), 8. Similar considerations are used as arguments for hidden variables by Honderich, T., *A Theory of Determinism*, Oxford (1988), pp. 322—5.
[84] Hutchison, K., 'What Happened to Occult Qualities in the Scientific Revolution?', *Isis* **73**, 233—53, especially 249—50 (1982).
[85] *Ibid.*, 250. One genuine difference between Pomponazzi and Hobbes is in their respective concepts of God. A crucial question here is the extent to which Hobbes' God was involved in the natural world. For an important argument that the God of Hobbes' writings on determinism is not the accepted Hobbist *deus absconditus*, see Pacchi, A., 'Hobbes and the Problem of God', in G. Rogers and A. Ryan, (eds.), *Perspectives on Thomas Hobbes*, Oxford (1988), pp. 171—87.
[86] Poppi, *op. cit.* (note 9), p. 666.
[87] For example, Bramhall on the 'horrid consequences' of determinism, which 'dishonours the nature of man', in Hobbes, *English Works, op. cit.* (note 59), V.III; cf. More, H., *The Immortality of the Soul* (1659), ed. A. Jacob, The Hague (1987), I.ix.1; Eliot, T., 'John Bramhall', in *Selected Essays*, London (1951), e.g. p. 358; Peters, R., *Hobbes*, London (1956), pp. 184—5.
[88] Bramhall, *op. cit.* (note 59), V. 21.
[89] Shapin, S. and Schaffer, S., *Leviathan and the Air-Pump: Hobbes, Boyle and the Experimental Life*, Princeton (1985), p. 104, note 64. Those who consider Shapin and Schaffer's Hobbist Hobbes to have little to do with the real Hobbes can treat the present remarks purely as part of a history of reception.
[90] See for example Dollimore, J., *op. cit.* (note 21), chapters 1, 10, 16; Barker, F., *The Tremulous Private Body: Essays on Subjection*, London (1984); Belsey, C., *The Subject of Tragedy*, London (1985), chapter 4. There is important material for a political history of the free will debate in Tyacke, N., 'Puritanism, Arminianism, and Counter-Revolution', in C. Russell, (ed.), *The Origins of the English Civil War*, London (1973), pp. 119—43.
[91] *Op. cit.* (note 72), chapters, pp. 110—33.
[92] Laney, B., 'Observations', appended to Hobbes, T., *A Letter About Liberty and Necessity*, London (1677), pp. 103—4.
[93] Milton, J., *Comus*, ed. W. Bell, London (1899), pp. 597—9.

[94] Quoted by Hunter, M., 'The Problem of Atheism in Early Modern England', *Transactions of the Royal Historical Society* **35**, 155 (1985).
[95] Bramhall, *op. cit.* (note 59), V. 25.
[96] For example, *English Works, op. cit.* (note 59), IV.252—8.
[97] *Ibid.*, III.297.
[98] Marston, J., *Antonio's Revenge, op. cit.* (note 23), IV.i.31—4; see Dollimore, J., *op. cit.* (note 21), pp. 36—40.
[99] *English Works, op. cit.* (note 59), IV.255—6.
[100] Compare the optimistic claim of a modern determinist, who thinks he is also a compatibilist, that we will keep all our social and moral practices in place even 'now that we see the social utility of the myth of free will'. Dennett, D., *Elbow Room*, Cambridge, Mass., and London (1984), chapter 7 and p. 166. Good modern accounts of the intuitions behind compatibilism and incompatibilism are Strawson, G., *Freedom and Belief*, Oxford (1986), pp. 105—20, and Honderich, T., *op. cit.* (note 83), pp. 382—400.
[101] *English Works, op. cit.* (note 59), IV.256—7.
[102] Bacon, F., *Works* 14 vols, (eds.) J. Spedding, R. Ellis, D. Heath, (1857—61), repr. Stuttgart (1961—3), XIII.213.
[103] *English Works, op. cit.* (note 59), V.55.
[104] Webster, *The Duchess of Malfi*, III.v.78—81.
[105] Sidney, P., *Arcadia* II, in *Works* (ed.) A. Feuillerat, Cambridge (1922), I.227.
[106] *The Duchess of Malfi*, V.iv.54—5.
[107] *English Works, op. cit.* (note 59), V.111.
[108] Plautus, *Captivi*, Prol. 22. *Fortune's Tennis* was the title of a play by Dekker, now lost.
[109] *Arcadia* V, *op. cit.* (note 105), II.177.
[110] Burton, R., *The Anatomy of Melancholy*, quoted by Trevor-Roper, H., *Renaissance Essays*, London (1985), p. 270.
[111] Traditional Webster criticism, either attacking his perverted and macabre sensationalism or somehow finding a deeply religious Christian humanism lurking beneath the plays' apparent anarchy, has only recently been challenged by readings which mark the sustained antihumanism of his thought. Particularly relevant here is Kroll, N., 'The Democritean Universe in Webster's *The White Devil*', *Comparative Drama* **7**, 3—21 (1973). She notes, among other things, the play's systematic materialism, which extends to the human mind, its emphasis on contact action between bodies, and its insistence that the universe, and life itself, is nothing but continual motion. Forker, C., 'The Tragic Indeterminacy of *The Duchess of Malfi*', chapter 7 in *Skull Beneath The Skin: The Achievement of John Webster*, Southern Illinois (1986), refers (pp. 367—8) to Pomponazzi's double truth, in which reason and faith are incompatible but complementary. But Forker's notion of a dual causality which preserves freedom, derived from the work of R. Grudin, on Paracelsus' influence, in *Mighty Opposites: Shakespeare and Renaissance Contrariety*, Los Angeles and London (1979), chapter 2, is an over-optimistic reading of Webster, even if it is applicable to Shakespeare. Pomponazzi's attitude to this 'double truth', too, is somewhat less than straightforward (section III above).
[112] Shapin and Schaffer, *op. cit.* (note 93), pp. 284—98. On censorship see Hill, C., 'Censorship and English Literature', in *The Collected Essays of Christopher Hill, Volume One: Writing and Revolution*, Brighton (1985), pp. 32—71.

[113] See now Malcolm, N., 'Hobbes and the Royal Society', in G. Rogers and A. Ryan (eds.), *op. cit.* (note 85), pp. 43—66.
[114] Lenoble, R., *Mersenne, ou la naissance du mécanisme*, Paris (1943), p. 133.
[115] Webster, C., *From Paracelsus to Newton*, Cambridge (1982), p. 89.
[116] Mersenne, *op. cit.* (note 79); *L'Impiété des Déistes*, Paris (1624).
[117] Hine, W., 'Mersenne, Naturalism and Magic', in B. Vickers, (ed.), *Occult and Scientific Mentalities in the Renaissance*, Cambridge (1984), pp. 165—76.
[118] *Op. cit.* (note 79), section 3.
[119] See note 53.
[120] Lenoble, *op. cit.* (note 114), p. 111.
[121] *Op. cit.* (note 79), col. 386.
[122] *Ibid.*, col. 1298; Lenoble, *op. cit.* (note 114), pp. 108—9, 300—2; cf. Hutchison, *op. cit.* (note 43), on other proponents of mechanism who thought that the very barrenness of matter on which their natural philosophy of contact action was based actually *guaranteed* the existence of a separate realm of the supernatural, including God, the mind, angels and demons, which intervenes in the otherwise passive material world. If there is no real distinction between natural and supernatural, it matters little, from Mersenne's point-of-view, whether God is immanent, as for Stoics and naturalists, or transcendent, as for radical supernaturalists. Mersenne *must* make a clean distinction between the mechanistic physical universe and the transcendent supernatural.

Liberty of indifference was a technical term of the Molinist Jesuits. It seems also to have been accepted by Descartes: see *Principles of Philosophy*, trans. V. Miller and R. Miller, Reidel (1983), I.x1, xli.
[123] Lenoble, *op. cit.* (note 114), pp. 571—2.
[124] *Ibid.*, p. 301.
[125] Dihle, *op. cit.* (note 6).
[126] Watson, G., 'Free Action and Free Will', *Mind* 96, 172 (1987).
[127] Webster, *The Duchess of Malfi*, III.v.81—3.
[128] Donne, J., 'An Anatomy of the World', pp. 155—7, in *The Complete English Poems*, (ed.) A. Smith, Harmondsworth (1971).
[129] Hooker, R., *Of the Laws of Ecclesiastical Polity*, with an introduction by C. Morris, 2 Vols., London (1969) III.viii.4.
[130] Cf. Calvin, *Institutes*, *op. cit.* (note 28), II.931: 'by His just and irreprehensible but incomprehensible judgement He has barred the door of life to those who He has given over to damnation'. See also Baker-Smith, D., 'Religion and John Webster', in (ed.) B. Morris, *John Webster*, London (1970), pp. 207—28.
[131] On this and other issues relevant to the free will debate in England see Tyacke, N., *op. cit.* (note 90); and Colie, R., *Light and Enlightenment: A Study of the Cambridge Platonists and the Dutch Arminians*, Cambridge (1957).
[132] Quoted by Porter, H., *Reformation and Reaction in Tudor Cambridge*, Cambridge (1958), p. 377. See Gunby, D., *The Duchess of Malfi*: A Theological Approach', in (ed.) B. Morris, *op. cit.* (note 130), pp. 179—204.
[133] Pilkington, J., *Works*, (ed.) J. Scholefield, Cambridge (1853), p. 227.
[134] Calvin, *Institutes*, *op. cit.* (note 28), I.xiv.20, 21.
[135] *The Duchess of Malfi*, III.v.69—71.
[136] Cudworth, R., Letter to Limborch, (1668), quoted by Cassirer, E., *The Platonic Renaissance in England*, London (1953), pp. 122—3.

JAMIE C. KASSLER

THE PARADOX OF POWER:
HOBBES AND STOIC NATURALISM*

> ... Love mankind. Walk in God's ways.
> 'All under law,' quoth the sage ...
> Marcus Aurelius
>
> To enter into these bonds is to be free
> John Donne, Elegy XIX

INTRODUCTION

In a brief passage at the beginning of *Leviathan*, Hobbes wrote about 'artificial life' by comparing the heart to a spring, the nerves to strings and the joints to wheels.[1] From this comparison some commentators suppose that Hobbes regarded living creatures as automata and that from such a mechanistic model his commonwealth was constructed. Other commentators, perhaps the majority, believe that Hobbes' mechanism is based on materialistic atomism of the Epicurean variety and that, accordingly, he was an advocate of egoistic individualism. When consideration is paid to the original cause of the motion of automata, commentators are widely divergent, some holding that Hobbes is an atheist and others, that Hobbes' God is the same as Aristotle's prime mover.[2]

If Hobbes were a neo-Epicurean, he would have developed the physical aspects of his philosophy by means of an emission theory, in which matter ('species') is actually transmitted. Hobbes did not do this in the writings of his middle and late period, for the physical theory propounded there is a continuum theory, transmission being that of a state rather than of matter. More importantly, Hobbes argued that only by means of a continuum does the world cohere. When discrete or individual parts make their appearance in Hobbes' writings, it is a sign of a lack of coherence, as in the case of dreams (or dreamlike states) where the self does not cohere and in the case of the state of nature where society does not cohere. Hobbes stated the first point clearly, while hinting the second, in the following passage from his least philosophical and most controversial work, *Leviathan*:

> ... Mentall Discourse, is of two sorts. The first is *Unguided, without Designe*, and inconstant; Wherein there is no Passionate Thought, to govern and direct those that follow, to it self, as the end and scope of some desire, or other passion: In which case the thoughts are said to wander, and seem impertinent one to another, as in a Dream. Such are Commonly the thoughts of men, that are not onely without company, but also without care of any thing; though even then their Thoughts are as busie as at other times, but without harmony; as the sound which a Lute out of tune would yeeld to any man; or in tune, to one that could not play.[3]

The confusion of commentators about Hobbes' God can be explained, I believe, by the reading that I propose in this paper, where I emphasise Hobbes' naturalism and his identification of God's will with creative activity in the universe. In particular, I shall argue that Hobbes is indebted to the main tenets of Stoicism, according to which God's will is a universal cause, immanent in everything and directing events by expansion and contraction to achieve worthy ends. If Hobbes is understood as a neo-Stoic, the problematic nature of Hobbes' God quickly comes into focus, for 'the Stoic conceived of the divine in terms that tended strongly towards monotheism, (but) he found it difficult to posit the existence of a God who was not in every way identical with the universe'.[4]

Hobbes' Stoicism was remarked on by at least one contemporary writer, John Bramhall (1594—1663), who charged him with adumbrating the Stoic conception of fate.[5] Hobbes answered Bramhall by drawing on the work of Justus Lipsius (1547—1606), a writer who asserted the congruence of Stoicism with Christianity.[6] Lipsius had divided destiny into four kinds: mathematical or astrological destiny, natural destiny, Stoical or violent destiny and the godly destiny of the Christians.[7] And Hobbes himself pointed out that Lipsius defined the last kind of destiny 'just as T. H. doth his destiny', as 'a series or order of causes depending upon the divine counsel'.[8] Lipsius' third kind of destiny, the Stoical or violent destiny, was a doctrine of conflagrations and regenerations of the universe, not a doctrine of fate. Rather, the Stoic doctrine of fate was that which Lipsius described, and Hobbes assented to, as the godly destiny of the Christians and which was conceived as a 'chain' of necessary causes.

This Stoic doctrine of causes provoked a great deal of controversy in its own time; and the most violent objectors were those who felt that determinism did away with human freedom and, hence, with human responsibility. Hobbes' philosophy provoked similar controversy; indeed,

Bramhall's charge focused specifically on the problem of freedom. The Stoics were not unaware of what a doctrine of fate could mean for morality, as Josiah B. Gould points out in his important study of the Stoic, Chrysippus.[9] This philosopher, according to Gould, tried to work out a deterministic doctrine and at the same time make a place for human responsibility. His solution was to distinguish between things that are 'in our power' and things that are not. Since he also recognized that maintaining both these strands at the same time was a logical impossibility, he appears to have held them successively or alternately.

In adopting a determinist position, Hobbes, too, had to confront the problem of human responsibility. Like Chrysippus, he distinguished between things that are in our power and things that are not. But he developed this distinction by acknowledging that what we will is determined not merely by external situations but also by our internal character. More importantly, he adopted a new definition of freedom. The customary definition of freedom was a privilege held by grant or prescription by which men enjoy some benefit beyond the ordinary subject, whereas Hobbes understood freedom in the Stoic sense as the power to do as one thinks fit unless restrained by law.[10] According to his definition, we have the possibility of determining our own actions by acquiring knowledge of what is permitted and what is not.[11]

In what follows, therefore, I treat Hobbes' solution in two parts by drawing on his distinction between sensation and conception, the former of which has a passive sense, the latter an active one.[12] In the first part I outline Hobbes' cosmic determinism with reference to his theory of light in order to show that sensation, a motion inward, is natural, and so is not in our power. In the second part I set forth Hobbes' notion of human freedom to show that conception, a motion outward, is artificial, and so is in our power. In each part I digress to compare Hobbes' ideas with the work of two of his friends, the physician, William Harvey (1578–1657), and the poet, Abraham Cowley (1618–1667).[13]

Except for the digressions on Harvey and Cowley, most of my evidence comes from one text, *Body*, the English version of *De corpore*.[14] Hobbes began work on this text in the 1640s, but it was not published until the mid 1650s (1655 in Latin, 1656 in English). *Body*, however, is not an isolated text, for it forms the first of three parts in all of which Hobbes presented the last systematic presentation of his entire philosophy.[15] The second part, on human nature, is a continuation of

Hobbes' naturalistic argument, whereas the third part, on the elements of law, is a conclusion drawn from the naturalistic premisses. Since the standard edition of these parts is problematic, I have cited them infrequently. Nevertheless, in constructing the argument in this paper, I have tried to avoid the criticism which Hobbes levelled at biblical interpreters, that

> ... it is not the bare Words, but the Scope of the writer that giveth the true light, by which any writing is to bee interpreted; and they that insist upon single Texts, without considering the main Designe, can derive no thing from them cleerly; but rather by casting atomes of Scripture, as dust before mens eyes, make every thing more obscure than it is; an ordinary artifice of those that seek not the truth, but their own advantage.[16]

Most commentators seem to agree that Hobbes' scope was political. But if Hobbes adumbrated a version of Stoicism, then the message of his philosophy must be ethical, namely, to live in harmony with nature. I believe this is *the* fundamental tenet of his entire philosophy, but the difficult phrase 'to live in harmony with nature' needs interpretation. This paper sets out to do this in the context of the problem of determinism and responsibility.

1. NOT IN OUR POWER

1.1. *Nature's Government: Motion Inward*

In 1644 a version of Hobbes' theory of light was published and presented as the seventh book of Marin Mersenne's *Universae geometriae, mixtaque, synopsis et bini refractionum demonstratum tractatus*. The theory is now referred to as *Tractatus opticus*, a title assigned when it was reprinted in Molesworth's edition of Hobbes' Latin writings.[17] According to Alan E. Shapiro, this work began the kinematic tradition in the continuum theory of light.[18] Although almost every optical writer of note in the 17th century was aware of it, some writers did not realize the work was by Hobbes, since the treatise was attributed to him in a *monitum* on p. 548.[19] For these details and for the following summary, I am indebted to Shapiro's excellent account of the 1644 theory and of kinematic optics generally, to which readers are referred for Hobbes' scientific contributions and for the distinctive differences between Hobbes' theory and that of Descartes.[20]

In the *Tractatus opticus* Hobbes argued that the sensation of light is produced by a vibratory motion propagated through an all-pervading medium of uniform density. In this brief statement of his theory, we have the two Stoic tenets of tension and goodness. First, we have tension conceived as vibratory motion, that is, as a real but insensibly small (infinitesimal) expansion and contraction. Hobbes compared this motion to that of the heart: like the systole and diastole of the heart, the entire body of the sun expands and contracts together, so that all the rays of light emanate radially from the center of the sun. To use Shapiro's words, the 'analogy implies a regularity but not a strict periodicity in the oscillations'.[21]

Second, we have goodness conceived as relief from ignorance ('darkness'), for the all-pervading medium, which Hobbes introduced as a 'crucial, unstated hypothesis', enables sensation to take place.[22] This is so, because, not only the luminous source but also the medium vibrates, that is, expands and contracts. In this way light is propagated through the medium by a vibratory motion to the retina and through the optic nerve to the brain, which by its reaction transmits the motion back again towards the sun. Only the motion propagated outwards by the reaction of the brain is called light, but it consists of a representation conceived in the brain. For conception to take place, therefore, two things are required: sensation, which is a motion directed from the external world inward to the sentient, and representation, which is a motion directed from within the sentient outward.

In *Body* Hobbes introduced three significant modifications to the 1644 theory.[23] First, he explained light by a 'simple circular motion' and not by expansion and contraction. Then, he altered the all-pervading medium from a uniform substance to a fluid composed of particles of different sizes and degrees of elasticity ('hardness'). Finally, he made the heart the 'fountain of sense' and the source of the reaction outwards. Let me take each of these modifications in turn and, first, the introduction of simple circular motion in which every line in a body is always moved parallel to itself. The problems of this modification for Hobbes' theory of light have been noted by Shapiro, who points out that such a motion cannot account for the sun's radiation, since light would only be radiated in directions normal to the axis of rotation and not in all directions. According to Shapiro, even though Hobbes was aware of this problem, he seems to have dropped the theory of expansion and contraction, because he thought it demanded the existence of a vacuum,[24] and in *Body* Hobbes denied the existence of a vacuum.[25]

But if Hobbes restricted the sun's external motion to simple circular motion, does the ambient medium (now a stated hypothesis) also move in a circular fashion? Hobbes' answer to this question is somewhat obscure and must be teased out by the few clues he provided as to the nature of the medium. These clues indicate that it is the same in all bodies; that it has no weight ('gravity'); that it fills up the universe so as to leave no empty space: that it is fluid in consistency; and that this fluid has no motion at all but what it receives from little bodies, that is, 'particles' or 'atoms' of different sizes that float in it and which are not themselves fluid.[26] These atoms, which have different degrees of hardness, are moved by the sun so that they perpetually change places one with another. But the sun does not effect this by its simple circular motion alone, for Hobbes pointed out that there also is a 'motive power' *in* the sun.[27] By means of the sun's compounded motions, then, homogeneous atoms are congregated and heterogeneous atoms are scattered, a process Hobbes called 'fermentation', from the Latin *ferveo*, meaning to boil, seethe, steam; to foam; to swarm; to be busy, bustle about; and, figuratively, to burn, glow, rage, rave.[28]

But Hobbes also indicated that 'the generation of the light of the sun is accompanied with the generation of heat',[29] and that 'when a body hath its parts so moved, that it sensibly both heats and shines at the same time, then it is that we say fire is generated'.[30] Both of these statements occur within the context of the theory of light, where Hobbes asserted that he who could account for 'whence, and from what action, both the *shining* and *heating* proceed, may be thought to have given a possible cause of the generation of *fire*'.[31] In proposing such a cause Hobbes argued that heat and pain are not in the fire but within us,[32] for 'when we grow hot, we find that our spirits and blood, and whatsoever is fluid within us, is called out from the internal to the external parts of our bodies, more or less, according to the degree of heat; and that our skin swelleth'.[33]

We now have the grounds for Hobbes' third modification, by which he made the heart the fountain of sense. Initially, Hobbes identified sense as the usual five — seeing, hearing, smelling, tasting and touching, the common organ of which is the brain.[34] But he afterwards identified 'another kind of sense', namely, pleasure and pain. This kind of sense proceeds

> ... not from the reaction of the heart outwards, but from the continual action from the outermost part of the organ towards the heart. For the original of life being in the heart,

that motion in the sentient, which is propagated to the heart, must necessarily make some alteration or diversion of vital motion, namely, by quickening or slackening, helping or hindering the same. Now when it helpeth, it is pleasure; and when it hindereth, it is pain...[35]

In this context vital motion is the motion of the blood 'perpetually circulating ... in the veins and arteries'.[36] When its circulation is helped, we feel pleasure; when it is hindered, we feel pain. But the common organ of this kind of sense is the heart.[37]

Hobbes, therefore, established a clear-cut separation between the function of heart and brain.[38] But these organs are merely instruments of the blood, the source and centre of life, for the blood ministers to, indeed activates, the motions of the heart and the brain, both of which have a propensity to vibrate.[39] Yet, even the blood functions as an instrument, for its purpose is to circulate the ambient ethereal fluid which, in humans, is a spirit in the blood.[40] In circulating with the blood, these spirits pass through the veins to the heart, where they are 'purified' and driven from the heart through the arteries. Upon reaching the brain, the spirits are impelled into, and retracted out of, the roots of the nerves at the base of the brain.[41]

1.2. *The Natural Life: Hobbes and Harvey*

All three modifications are illuminated by recourse to the work of William Harvey, whose major achievement, as Hobbes recognized, was the discovery of the circulation of the blood.[42] Harvey announced this discovery in *De motu cordis*,[43] where he compared the blood's circulation to the motion produced in distillation of the water cycle. Transformed by distillation (a chemical concept), the blood becomes an example of antiperistasis — whereby all movement in space is circular thrust — in which substances succeed departing substances, thereby preventing the formation of a vacuum.[44] In that same work Harvey argued that the source of life was heat, namely, the heart as a unit of flesh, fibre, blood and spirit. Accordingly, he assigned to the heart the chief part in the body.[45] In other writings he likened the action of the heart to a pump, the force of which drove the blood round in a circle.[46]

Since Harvey also regarded the heart as a muscle and, hence, passive, the pump analogy did not settle the question of how, in the embryo, the heart first beats.[47] In addressing himself to this question,

Harvey changed his mind about the role of the heart. In *De generatione animalium* he argued that the blood, not the heart, was the source of life and co-extensive with heat, that is, innate spirit.[48] Hence, the hottest part of the body was the blood, which imparted heat to the heart rather than received heat from the heart. According to this modification, Harvey supposed that, in the embryo, the heart's first diastole was a 'fermentation', the origin of which was heat in the embryonic blood.[49]

But if fermentation served only to ignite the heart beat, what was responsible for the continued beating of the heart throughout life? The answer to this question is to be found in the few hints Harvey provided about an 'obscure' pulsating motion inherent in the blood.[50] Indeed, Harvey supposed that as the blood is the 'first born', so 'in it and from it pulsation begins'.[51] But he also stated that when the heart ceases beating in a dying animal, 'you will perceive in the blood itself a kind of undulation and vague fluttering or palpitation, the last token of life'.[52] On the basis of this internal motion, Harvey inferred that the blood was a living creature and, therefore, sentient, for from variations in the blood's pulsations, it was plain, Harvey wrote, 'how sensitive it is to harm done to it by things that are hurtful and to the comfort of things that cherish it'.[53]

That sensation and movement exist before the brain develops in the embryo is a paradox, as Harvey recognized.[54] To resolve this paradox, Harvey added to the usual five senses another sense: the feeling of pleasure and pain, a feeling that was different from the sense of touch.[55] He also distinguished different functions for heart and brain. According to him, the heart governs all natural motions and actions which 'go on whether we will or no', whereas the brain governs all animal motions.[56] But these 'instruments' perform their 'public offices' because of the circulation of the blood.

When circulation is helped, heart and brain function normally, according to their inherent 'harmony and rhythm'. But when the circulation is hindered, neither heart nor brain function properly, since variations in the blood's pulsation bring disorder and, consequently, disease.[57] In *De generatione animalium*, therefore, Harvey centralized all physiological functions in the blood: the blood flows round the whole body, imparting heat and life continually to all its parts. Hence, Harvey concluded that 'the soul residing in it primarily and principally may justly be thought to exist in it for the blood's sake, and to be altogether in the whole and altogether in each part'.[58]

Here, then, we have the source of all three modifications which Hobbes made to his 1644 theory of light. But in converting Harvey's natural animal into cosmic animal, Hobbes changed at least one aspect of Harvey's physiology, for Harvey had consistently denied the particulate nature of living blood: it became so only as gore.[59] The reasons why Hobbes adopted a particulate fluid as the cosmic equivalent of the blood must remain conjectural, but I believe they are to be sought in a problem which had puzzled Harvey.

In *De generatione animalium*, and against prevailing opinion (another paradox), Harvey had reintroduced Aristotle's doctrine of epigenesis. According to Harvey's account, all animals have a uniform pattern of generation which begins when the male semen activates the egg by exerting some vital influence. But, lacking a microscope, Harvey was prevented from seeing the male spermatozoa, so he was unable to point to a contribution from both sexes.[60] Consequently, he supposed that fertilization could be explained by the analogy of contagion, that is, by 'some kind of contact' or 'touch'.[61]

But Harvey went farther and supposed that the 'fecundating contagion or first conception' is

... like the conception of the brain, so that fecundity be acquired in the same way as knowledge (for there is no lack of arguments to prove it), and whether, like the movements and animal operations which take their origin from the conception of the brain and we call appetites, natural movements also and the operations of the vegetative faculty (especially generation) depend upon the conception of the womb.[62]

Although Harvey himself never seems to have settled on an explanation of contagion, he suggested that the problem would be 'rightly and piously' investigated by one 'who deduces the generation of all creatures from that same eternal and omnipotent Deity upon whose nod the whole universe itself depends'.[63] On this assumption, an explanation of contagion would have to account for how, in a chain of efficient causes, the 'final' efficient cause acts at a distance.

At the conclusion of *Body* Hobbes provided an hypothesis of how this action might take place, when he wrote about magnetism as follows: 'It is ... certain, that the attractive power of the loadstone is nothing else but some motion of the smallest particles thereof' and that this motion is 'reciprocal motion in a line too short to be seen'.[64] Hobbes then continued:

> Now in what manner and in what order of working this cause produceth the effect of attraction, is the thing to be enquired. And first we know, that when the string of a lute or viol is stricken, the vibration, that is, the reciprocal motion of that string in the same strait line, causeth like vibration in another string which hath like tension.[65]

For both Hobbes and Harvey, therefore, the first link in the chain of causes is touch; for both men, a fluid substance is the instrument of touch, since it transmits the creative power of nature. While Harvey remained puzzled about the manner in which this power was transmitted, Hobbes provided an hypothesis in his analogy to sympathetic resonance. The implications of this analogy are twofold: first, that the ambient fluid has a dynamic property similar to a tense string; second, that affinity is a pre-condition of vibration.

2. IN OUR POWER

2.1. *Self-Government: Motion Outward*

From the evidence adduced so far, it is apparent that Hobbes' explanation of nature's power is a development of Stoic naturalism, according to which the world is a continuum filled with an all-pervading *pneuma* [Lat.: *spiritus*], which serves to generate sympathy [*sympatheia*], the cohesion of matter and contact between all parts of the cosmos. But cohesion — the interaction and affinity of different parts of a unified structure — is maintained by tension [*tonos*], a dynamic property of the all-pervading *pneuma*. Central to this notion is the seed, for it is the seed that is *pneuma*, and the character of this seed is heat and, hence, active force.[66] Since modifications of *pneuma* take place in the body as modifications of tension, cohesion refers to a chain of efficient causes on a cosmic scale.[67]

There are three points to note about Stoic naturalism. First, the plenum necessitates circular motion of bodies. This is the ideal motion, a mark of totality and perfection. Thus, Marcus Aurelius observed that the 'soul attains her perfectly rounded form when she is neither straining out after something nor shrinking back into herself'.[68] Second, circular motion may be altered by modifications of *pneuma*. But, and this is the third point, for motion to take place at all there must be contact between bodies and *pneuma*. This contact is achieved by tensional motion [*tonike kinesis*], the propagation of an impulse through a medium in a state of tension.[69] Regulatory power, therefore, is given to the *pneuma* and not, as in the Aristotelian cosmos, to the circular movements of the celestial spheres.

I suggest that Hobbes' emendations to the 1644 theory of light were due principally to his more global aims in *Body*. According to the cosmic determinism elaborated there, the world is a fluid plenum filled with atoms of matter.[70] But matter has two principles, a passive and an active principle. The manifold variety in the world is explained by reference to these two features of matter.[71] The passive principle is unqualified body; the active principle is heat or innate spirit, that is a seething motion of atoms in the ambient fluid. I shall now call this motion 'impulse', because it functions as a stimulus to excite ('help') or inhibit ('hinder') other, grosser motion in the universe. Hence, impulse is the source for all activity, change and variety and, therefore, for knowing.

In humans, impulse is the stimulus for exciting or inhibiting vital and animal motion. Vital motion, being natural, is not in our power, for it is born with us and continues until death. Such are the motions of the blood, pulse, breathing, concoction, nutrition and excretion.[72] At a certain level of biological organization, impulse also generates animal motions, the first appearance of which Hobbes identified with 'the Interiour Beginnings of Voluntary Motions'.[73] Such motions grow with us,[74] for they are acquired in the same way as a musician acquires musical skills, by habituation.[75] Animal motions, therefore, are artificial and, so, are in our power.

But animal motions involve two things, namely, deliberative processes, or what Hobbes called 'reckoning',[76] and the use of our limbs, for the *product* of animal motions are the passions, the object of which is some sensible good.[77] But sensible good and virtue are not the same, for only the right use of the will entails the virtuous life. Unlike the passions, the object of the will is that good which reason leads us to seek. It is not in our power not to be stirred mentally by our passions; but it is in our power to translate them or not to translate them into action.[78] As Hobbes wrote in *Body*,

Neither is the freedom of willing or not willing, greater in man, than in other living creatures. For where there is appetite, the entire cause of appetite hath preceded; and, consequently, the act of appetite could not chose but follow, that is, hath of necessity followed ... And therefore such a liberty as is free from necessity, is not to be found in the will either of men or beasts. But if by liberty we understand the faculty or power, not of willing, but of doing what they will, then certainly that liberty is to be allowed to both, and both may equally have it, whensoever it is to he had.[79]

Passion, then, is the will's solicitor, whereas the will is passion's controller. But it is only by being thoroughly enlightened by reason that the will can be victorious. Though it is possible to make a wrong choice through an error of judgment — what Hobbes called 'misreckoning', it is also possible for the will to be so corrupt as to go against the evidence of reason.[80]

In Hobbes' account of cognition, the lowest form of reason concerns our immediate acts only. Will, in this context, is practical reason, for it is the outcome of prudential calculations and generalizations concerning the utility of a given act.[81] But practical reason is an attribute we share with other animals.[82] To become fully human, therefore, we must aspire to higher forms of knowledge. One form is knowledge of God's will, and this is the same as knowledge of nature; the other form is self-knowledge, that is, knowledge of one's own will. Thus, at the outset of *Leviathan* Hobbes argued:

> ... there is a saying much usurped of late, That *Wisedome* is acquired, not by reading of *Books*, but of *Men*. Consequently whereunto, those persons, that for the most part can give no other proof of being wise, take great delight to shew what they think they have read in men, by uncharitable censures of one another behind their backs. But there is another saying not of late understood, by which they might learn truly to read one another, if they would take the pains; and this is, *Nosce teipsum, Read thy self*...[83]

Accordingly, self-knowledge is not egoism; rather, it is the beginning of all virtue.

To account for virtue, Hobbes had recourse to the phenomenon of tension, whereby virtue or its opposite arises from the strength or weakness, tautness or slackness, of the self. The pre-condition for virtue, therefore, is affinity of the various parts of the self, all of which together form a unity.[84] If there is no resonance or cohesion of the self, we behave as if asleep or as in a dream or as if distempered with sickness.[85] The self, therefore, requires tempering in the same way that the strings of a musical instrument require adjustment to bring them into agreement. Indeed, in matters pertaining to the will, the chief virtue is temperance, that habit in choice and avoidance which preserves the judgments of reason. Temperance, therefore, may be defined as the exercise of tension in matters of choice.[86] And this demands consideration of the role of the passions in moral action.

Although passion is co-substantial with nature, the individual has an identity in his own right, since human nature possesses many passions, differing with the individual.[87] What, then, is the best condition of the self? Hobbes' answer to this question is completely different from that of Aristotle. As Theodore J. Tracy demonstrates, Aristotle's model of the self was the lever, the fulcrum being the fixed center (the prime mover/heart), the unmoving point from which opposite motions arise.[88] For Hobbes, the self is like a taut string: if the string is too tense, it will break; if it is too slack, there will be no 'passion'.[89] The best condition, therefore, is that degree of inner tension which checks the passions and brings them into harmony. Failing this, the self will be overcome by the power of excessive passion, and so pushed toward vice or madness, both of which are contrary to reason.[90] To be properly free, and therefore wise, a person must liberate himself from the vices that derive from excessive passion.

2.2. *The Happy Life: Hobbes and Cowley*

That human freedom consists in self-restraint was asserted by Abraham Cowley in his collection entitled *Several Discourses by Way of Essays, in Verse and Prose*,[91] which commences as follows:

> The Liberty of a people consists in being governed by Laws which they have made themselves, under whatsoever form it be of Government. The Liberty of a private man in being Master of his own Time and Actions, as far as may consist with the Laws of God and of his Country.[92]

In this collection Cowley presented one sustained philosophic argument that centres on the Stoic paradox concerning the Great Man and the Happy Man: while the former, in having all, has nothing, the latter, in having nothing, has all. Against the Puritan creed of a highly emotional conversion and subsequent daily exposure in the eternal battle between good and evil, Cowley pitted the Stoic exhortation never to give in to passion or fanaticism but always to preserve a soundly balanced spirit untouched by the vicissitudes of life. Happiness is a question of internal peace: the happy man is he who, having nothing, yet has all, because he is completely self-possessed and serene.[93]

The passions that enslave Great Men are outward signs of vexation of spirit, so that Cowley only 'slightly' touched upon

... particulars of the slavery of Greatness: I shake but a few of their outward Chains; their Anger, Hatred, Jealousie, Fear, Envy, Grief, and all the *Etcaetera* of their Passions, which are the secret, but constant Tyrants and Torturers of their life, I omit here, because though they be symptomes most frequent and violent in this Disease; yet they are common too in some degree to the Epidemical Disease of Life it self.[94]

In developing the notion of freedom as self-restraint, Cowley employed a symbol that may be traced back to Hellenic times.

Why, I'le tell you who is that true Freeman ... ; Not he who blindly follows all his pleasures (the very name of Follower is servile) but he who rationally guides them, and is not hindred by outward impediments in the conduct and enjoyment of them. If I want skill or force to restrain the Beast that I ride upon though I bought it, and call it my own, yet in truth of the matter I am at that time rather his Man, then he my Horse.[95]

Here, the unruly horse represents Cowley's own passions, and mastery of the horse signifies mastery of virtue over vice.

In addition to what Cowley called 'his Man', we are enslaved by other men, as well as by 'Custom, Business, Crowds, and formal Decency'.[96] To symbolize this more inclusive tyrant, Cowley replaced the horse with a beast drawn from Scriptures, when he outlined the wretched existence of a Great Man, 'guarded with Crowds and shackled with Formalities'.

The half hat, the whole hat, the half smile, the whole smile, the nod, the embrace, the Positive parting with a little Bow, the Comparative at the middle of the room, the Superlative at the door; and if the person be *Pan huper sebastus*, there's a *Huper-superlative* ceremony then of conducting him to the bottom of the stairs, or to the very gate: as if there were such Rules set to these *Leviathans* as are to the Sea, *Hitherto shalt thou go, and no further*.... Thus wretchedly the precious day is lost.[97]

In this passage the over-elaborate social code is the Leviathan, because its rules fetter not merely the Great Man but also his suitors.

According to Cowley, happiness cannot be found in the kind of servitude that court life requires. Indeed, in his essay, 'Of My self', Cowley intimated that contentment — internal peace — is unobtainable, since '... God laughs at a Man, who says to his Soul, *Take thy ease*'.[98] Nevertheless, by nature we have a tendency to seek internal peace, and our endeavour to obtain this unobtainable goal is what Cowley called 'the Epidemical Disease of Life'.[99] Here, then, we find the fundamental assumption of Stoicism: that all life is a striving, a belief clearly stated by Hobbes in *Leviathan*, where he wrote: 'there is no such thing as perpetuall Tranquillity of mind, while we live here; because Life it selfe is but Motion, and can never be without Desire, nor without Feare, no more than without Sense'.[100]

For Cowley, who had served at the court of the Stuarts, internal peace was to be sought in retirement.[101] But retirement is not the same as inactivity, for Cowley identified two ways in which to seek the elusive contentment, one, philosophical, the other, agricultural. Since few had the intellectual capacity for philosophical activity, Cowley settled for becoming a good husbandman. Nevertheless, he bracketed together the philosopher and the husbandman, because they share the same Stoic fortitude: they are self-governed, and so not playthings of fickle fate or superstitious fears.

Cowley's friend, Hobbes, chose the other way: to become a good philosopher. The emphasis, of course, must be on the word 'become', for Hobbes' philosophical writings are not mere repetitions one of another but contain developments and emendations. Nevertheless, they exhibit a consistent belief that the chief object of human striving is relief from ignorance ('darkness').[102] This *goal* is Hobbes' highest good; and it derives from his principle that the evident is true: since falsehood is the evil of the intellect, the complete goodness of the universe would have to be denied if the human desire to know were eternally thwarted.[103] Thus, he argued that 'neither things, nor imaginations of things, can be said to be false, seeing they are truly what they are; nor do they, as signs, promise any thing which they do not perform; for they indeed do not promise at all, but we from them'.[104]

In this extraordinary passage Hobbes rejected the classical view that nature manifests itself. For him, knowledge of nature is conditioned by thought and mediated by speech or by writing. This, too, was the view of the Stoics, who regarded speech as a symbol involving not only a sign and a thing signified but also a perceiver who makes a connection between the two.[105] Or, to put it in Hobbes' terms, speech or writing involves external reality, conceptions ('phantasms'), and our propositions and arguments, opposites and paradoxes. Hence, 'significant speech' always implies a linking between word and reality, so that through language can be shown how people re-arrange, combine and contrast their conceptions.[106] Accordingly, Hobbes identified two sources of error.

The first source, which Hobbes called 'affirming and denying', is an error of speech or of writing, in which error arises from misnomers as well as from hasty inference.[107] The second source, which he called 'perception and silent cogitation', is an error of thought, in which errors are tacit and are made in three ways:

> ... by passing from one imagination to the imagination of another different thing; or by feigning that to be past, or future, which never was, nor ever shall be; ... lastly, when from any sign we vainly imagine something to be signified, which is sense; and yet the deception proceeds neither from our senses, nor from the things we perceive; but from ourselves while we feign such things as are but mere images to be something more than images.[108]

Hobbes' two sources of error, then, are either false propositions or false conceptions, both of which he called by the name 'passions'. In his philosophy, therefore, the passions are false judgments.

According to Hobbes, reason must supervene on the passions, for he wrote:

> The best way ... to free ourselves from such errors as arise from natural signs, is first of all, before we begin to reason concerning such conjectural things, to suppose ourselves ignorant, and then to make use of our ratiocination; whereas, errors which consist in affirmation and negation (that is, the falsity of propositions) proceed only from reasoning amiss.[109]

The implication is that we need an appropriate method which may guide us in our generalizations and interpretations of nature's signs. In *Body*, Hobbes provided an exemplar of this method, which constitutes his own personal striving to reach a rational, that is, causal, understanding of the universe by viewing the parts in light of the whole and, so grasping the underlying principle.[110]

CONCLUSION

According to the argument in the foregoing pages, Hobbes' solution to the paradox of determinism and responsibility involves a complex of ideas relating virtue to knowledge and to imitation of the divine will. This complex of ideas begins with the dynamic power of nature which operates immanently by means of impulse, of which there are many degrees. All conative states in the cosmos, including in humans, are impulses, that is, movements toward or away from some thing. By these means natural phenomena are generated as well as corrupted ('dissolved'). For Hobbes, therefore, nature's power, which we may now call '*logos*', binds together every particular reality and event and assigns to them their unchanging measure.

Logos is not brought in from outside the biological realm, but at a certain level of organization arises spontaneously within it; and the form taken by impulse below this level is not irrational but prior to the rational.[111] People alone have the power of regulating this kind of impulse — passion — in accordance with *logos*; but when they fail to exercise this power, impulse is liable to become incommensurate and thus contravene the *logos* of universal nature, which impulse cannot do in any other creature. Hobbes' digressions about bees and ants must be understood in this context, for the impulses of these creatures do not, indeed, cannot, contravene universal nature.[112] *Logos*, therefore, is God's will; and it is here that we find Hobbes' moral imperative, because 'God knoweth the heart'.[113]

But God does not compel us to follow universal nature, for we have the possibility of determining our own actions by self-restraint. How do we accomplish this? Hobbes' answer is that the self needs to be slackened or tightened by philosophical education.[114] This is so, because in human society the deliberative process Hobbes called 'practical reason' produces a clash of wills and leads to social discord. For social concord another kind of deliberative process is necessary, one that involves counsel and study methodically pursued.[115] By this method we may cultivate our true self, one that is in sympathy not only with our own latent powers but also with the active powers of nature. By this method also we may achieve 'right reason' or 'evidence of truth',[116] the only kind of knowledge that entails the virtuous life.

When considering human actions, which are our will, *logos* becomes a norm in conformity with which people ought to mould their lives. Hobbes detailed this norm as a set of natural laws which in *Leviathan* he treated as 'dictates of Reason'. These dictates, he wrote,

> ... men use to call by the name of Lawes; but improperly: for they are Conclusions, or Theoremes concerning what conduceth to the conservation and defence of themselves; whereas Law, properly is the word of him, that by right hath command over others. But yet if we consider the same Theoremes, as delivered in the word of God, that by right commandeth all things; then are they properly called Lawes.[117]

Conformity to law 'binds' people together, cohesively, just as *logos* binds the universe together, cohesively. The virtuous life, therefore, is truly in our power, for it is a striving to live in conformity with laws we have made ourselves.[118]

NOTES

[*] The first draft of this paper was given at the Hobbes 1588—1988 Research Symposium, organized by Conal Condren and held at the University of New South Wales 9—10 July 1988. Thanks are due to Warren D. Anderson and Alan E. Shapiro for comments which have been taken into account when revising the paper. Upon completion of the final revision, a book was published that seemed to treat a theme common to my own: Herbert, G. B., *Thomas Hobbes: The Unity of Scientific and Moral Wisdom*, Vancouver (1989). Herbert's interpretation, however, is very different from mine, for he makes Hobbes a precursor to Leibniz and Hegel. Since much of Herbert's argument rests on his understanding of *conatus*, it is significant that he failed to consult the seminal work by Shapiro (see n. 18 below), who clarifies Hobbes' insight that a body and a pulse require different physical explanations.

NB: In those cases where I have had to rely on the Molesworth edition, Hobbes' writings are identified first by short title and then by the abbreviations HLW and HEW. These abbreviations refer respectively to (1) *Thomae Hobbes Malmesburiensis opera philosophica quae latine scripsit* ed. William Molesworth, 5 Vols, London (1839—45) and (2) *The English Works of Thomas Hobbes of Malmsbury*; now first collected and edited by Sir William Molesworth, Bart., 11 Vols, London (1839—45), reprinted, Scientia Aalen (1962).

[1] Hobbes, T., *Leviathan* [1651] edited with an introduction by C. B. Macpherson, Harmondsworth (1968), p. 81.
[2] For the various positions of different 'Hobbists', see Sacksteder, W., *Hobbes Studies (1879—1979): A Bibliography*, Bowling Green, Ohio (1982). For a recent study that assumes Hobbes' deity is *deus absconditus*, see Shapin, S. and Schaffer, S., *Leviathan and the Air-Pump: Hobbes, Boyle, and the Experimental Life* . . . Princeton (1985).
[3] Hobbes, *Leviathan*, p. 95.
[4] Anderson, W. D., *Matthew Arnold and The Classical Tradition*, Ann Arbor (1965), p. 133. My reading of Hobbes has been influenced by Roger North (1651?—1734), a neo-Stoic critic of Hobbes. See Kassler, J. C. 'Man a la Mode: or, Re-interpreting the Universe from a Musical Point of View', in Kassler, J. C. (ed.), *Metaphor — A Musical Dimension*, Sydney (1991) and sources cited there.
[5] For the controversy with Bramhall, see Hobbes, T., *An Answer to a Book . . . by Bramhall*, HEW 4: pp. 279—408 and *The Questions concerning Liberty, Necessity and Chance*, HEW 5.
[6] Lipsius helped to place Stoic ideas before a wide audience at the turn of the seventeenth century, according to Barker, P. and Goldstein, B. R., 'Is Seventeenth Century Physics Indebted to the Stoics?', *Centaurus* 27, 148—64 (1984).
[7] Hobbes, T., *The Questions concerning Liberty, Necessity and Chance*, HEW 5: pp. 242—3.
[8] *Ibid.*, HEW 5: p. 243.
[9] Gould, J. B., *The Philosophy of Chrysippus*, Leiden (1970), p. 152 and 'The Stoic Conception of Fate', *Journal of the History of Ideas* 35, 17—32 (1974).

[10] For the changing relationship of political power, property and law, see Wilson, C., *England's Apprenticeship 1603—1763*, London (1965), p. 9 *et passim*.
[11] For the Stoic background, see Long, A. A., 'Freedom and Determinism in the Stoic Theory of Human Action', in Long, A. A. (ed.), *Problems in Stoicism*, London (1971), pp. 173—99.
[12] Conception, in Hobbes' sense, equals the Stoic apprehension, for which see Sandbach, F. H. 'Phantasia Kataleptike', in Long, A. A. (ed.), *Problems in Stoicism*, London (1971), pp. 9—21.
[13] In 1657, the year of Harvey's death, Cowley was created M.D. at Oxford, which was to become the leading centre for the study of physiology in the second half of the seventeenth century. Ralph Bathurst (1620—1704) was one of Hobbes' contacts there. See Frank, R. G., *Harvey and the Oxford Physiologists: Scientific Ideas and Social Interaction*, Berkeley, Los Angeles, London (1980).
[14] *Elements of Philosophy. The First Section, Concerning Body*, HEW 1, hereafter cited as *Body; De corpore*, HLW 1.
[15] The other two parts of the English version are *Human Nature*, HEW 4: pp. 1—76 and *De corpore politico: or the Elements of Law*, HEW 4: pp. 77—228.
[16] Hobbes, *Leviathan*, p. 626.
[17] *Tractatus opticus*, HLW 5: pp. 215—48.
[18] Shapiro, A. E., 'Kinematic Optics: A Study of the Wave Theory of Light in the Seventeenth Century', *Archive for History of Exact Sciences* 11, 143—72 (1973).
[19] *Ibid.*, pp. 181—8.
[20] For the contact between the two men, see Brandt, F., *Thomas Hobbes' Mechanical Conception of Nature*, Copenhagen and London (1928).
[21] Shapiro, *op. cit.* (n. 18) p. 147.
[22] *Ibid.*, p. 146.
[23] For Hobbes' last theory of light, see *Body*, HEW 1: pp. 28, 75—9, 274—5, 286, 374—6, 378, 381—5, 404, 448—65, 497—8.
[24] Shapiro, *op. cit.* (n. 18), p. 169 and n. 116.
[25] See *Body*, HEW 1: p. 426. Hobbes' supposition of continuous atomism (a plenum) comes after his examination and rejection of the discrete atomism (a vacuum) of the Epicureans.
[26] *Body*, HEW 1: pp. 426, 448, 474, 481, 504, 519. Hobbes first mentioned a medium on p. 215, where he distinguished between a fluid and a consistent medium. By the latter he denoted 'a *medium* whose parts are by some power so *consistent* and *cohering*, that no part of the same will yield to the movent, unless the whole yield also'. And at p. 426 he indicated that there are degrees of coherence (consistency, tenacity); that is, a scale of degrees of rarity and density. This idea was present in the earlier *Tractatus opticus*, since, according to Shapiro, *op. cit.* (n. 18), pp. 154—5, Hobbes was 'the first to get the so-called "standard" velocity condition for a continuum theory of light, namely that the velocity of propagation is greater in a rarer than in a denser medium (*e.g.*, it is greater in air than water)'. For Hobbes' references to media, see *Body*, HEW 1: pp. 217, 321—4, 326—32, 334—5, 337—9, 341—2, 344, 374—83, 425—6, 509.
[27] *Body*, HEW 1: p. 430: 'by supposing motive power *in* the sun, we suppose motion also; for power to move without motion is no power at all' [italics mine].
[28] Ferveo/fermentation: *Body*, HEW 1: pp. 324—5, ?331, 449, 450, 474; *De corpore*, HLW 1: pp. 264, 336, 366, 392.

[29] *Body*, HEW 1: p. 448.
[30] *Body*, HEW 1: p. 449.
[31] *Body*, HEW 1: p. 451.
[32] *Body*, HEW 1: p. 449; for the implications of this theory, see *Human Nature*, HEW 4: pp. 6–8, and n. 76 below.
[33] *Body*, HEW 1: p. 449.
[34] *Body*, HEW 1: p. 403; *Human Nature*, HEW 4: p. 54.
[35] *Body*, HEW 1: p. 406.
[36] *Body*, HEW 1: p. 407.
[37] *Body*, HEW 1: p. 406.
[38] For the different functions of heart and brain, see *Body*, HEW 1: pp. 392–404, 448 and *Human Nature*, HEW 4: pp. 3–8, 10–1, 31, 34.
[39] *Body*, HEW 1: pp. 392, 486. According to Hobbes, all animal parts have a propensity to vibrate; moreover, memory consists of vibrations of the *pia mater* (*ibid.*, HEW 1: pp. 400–1). Cf. Harvey, W., *The Anatomical Lectures* ... edited ... by G. Whitteridge, Edinburgh and London (1964), pp. 309, 315. Harvey wrote the lectures between 1616 and c. 1618, but they remained unpublished until the 20th century.
[40] Spirits: *Body*, HEW 1: pp. 393, 397, 403, 407, 408, 449, 466, 505; *Human Nature*, HEW 4: pp. 4, 6, 11, 55, 56, 60–1, 63, 65.
[41] *Body*, HEW 1: pp. 397, 403, 408.
[42] *Body*, HEW 1: pp. viii–ix, 407. For aspects of the development of Harvey's thought, see Witteridge, G., *William Harvey and the Circulation of the Blood*, London and New York (1971). For the impact of Harvey's ideas on the development of physiology after c. 1650, see Frank, *op. cit.* (n. 13); Davis, A. B., *Circulation Physiology and Medical Chemistry in England 1650–1680*, Lawrence, Kansas (1973); and Mendelsohn, E., *Heat and Life: The Development of the Theory of Animal Heat*, Cambridge, Mass. (1964).
[43] The first edition, in Latin, was published in 1628; a revised version appeared in 1648; an English translation was published in 1653, reprinted 1673, and is the basis of Harvey, W. *An anatomical Disputation concerning the Movement of the Heart and Blood in living Creatures* translated ... by G. Whitteridge, Oxford (1976), hereafter referred to as *De motu cordis*.
[44] *De motu cordis*, pp. 75–6. For Hobbes' version of the water cycle, see *Body*, HEW 1: pp. 468–9.
[45] *De motu cordis*, p. 76.
[46] See Harvey, *The Anatomical Lectures* (n. 39), p. 273 and Franklin, K. J. (tr.), *The Circulation of the Blood: Two Anatomical Essays by William Harvey* ... , Oxford (1958), p. 60, hereafter referred to as *Letters to Riolan*.
[47] For his treatment of muscles, see Harvey, W., *De motv locali animalivm 1627* edited ... by G. Whitteridge, Cambridge (1959).
[48] Harvey began to change his mind about the heart in the 1630s; see Harvey, W., *Disputations touching the Generation of Animals* translated ... by G. Whitteridge, Oxford (1981), 111 *et passim.*, hereafter cited as *De generatione animalium*. Harvey commenced writing this manuscript in the 1630s and completed it by 1647–48. The Latin text was published in 1651, and an English translation appeared in 1653. Frank, *op. cit.* (n. 13), pp. 36–7, provides a tentative dating of the contents.
[49] Fermentation: *De generatione animalium*, p. 242; see also pp. 243, 257, 448, 463, 465.

⁵⁰ For example, *De motu cordis*, pp. 44—5, 47.
⁵¹ *De generatione animalium*, p. 241.
⁵² *De generatione animalium*, p. 243.
⁵³ *De generatione animalium*, p. 247, where the variations are stated to be the blood's 'speed or slowness, its vehemency or weakness etc.'. The blood, therefore, has an internal, pulsating motion as well as external circular movement, so that Harvey, *ibid.* 217, could write that the blood 'moves and dances like an animal'.
⁵⁴ *De generatione animalium*, pp. 293—300.
⁵⁵ *De generatione animalium*, pp. 298—9.
⁵⁶ *De generatione animalium*, pp. 296—9.
⁵⁷ *De generatione animalium*, pp. 247—8. For 'harmony and rhythm' see *ibid.* pp. 97, 100, 250, 299, 297 and *De motu cordis*, pp. 39, 50, 51.
⁵⁸ *De generatione animalium*, p. 245 (see also pp. 376, 382).
⁵⁹ For Harvey's conception of gore, see *Letters to Riolan*, pp. 38, 65; and *De generatione animalium*, pp. 243, 253—8, 374, 381.
⁶⁰ Twenty years after Harvey's death, Anton van Leeuwenhoek (1632—1723) saw spermatozoa for the first time with the aid of a microscope.
⁶¹ *De generatione animalium*, pp. 147—8, 183, 189—90, 227—8, 230, 232, 239—40, 353, 357, 443, 448, 463—5. For Hobbes' theory of contagion, see *Decameron physiologicum*, HEW 7: pp. 129, 136.
⁶² *De generatione animalium*, p. 239 (see also pp. 349—50, 351, 445—6, 452—3). The conception of the brain as an embryo is implicit in the terms '*dura mater*' and '*pia mater*' (derived from the Arabic), according to Onians, R. B., *The Origins of European Thought about the Body, the Mind, the Soul, the World, Time, and Fate* . . . , 2d edn., Cambridge (1951), p. 111 n. 5. See also Pagel, W., *William Harvey's Biological Ideas: Selected Aspects and Historical Background*, Basel and New York (1967), pp. 270—6 *et passim*.
⁶³ *De generatione animalium*, p. 237. A probable source for Harvey's notion, that the first cause of things is a 'nod' of the deity, may be Cicero, who used nod (L. *nutus*) in its figurative sense on a number of occasions: for example *Catilinarians*, 3.9. 21, *omnia deorum nutu atque potestate administrari* (all things are governed by the will and the power of the gods). In *De motu cordis*, p. 127, Harvey equated the term 'nod' with contraction, writing that

> . . . it is certain that all local movement in animals comes first and takes it beginning from the contraction of some particle. . . . This truth concerning local movement, and that the immediate motive organ in every movement of all animals in which there is from the beginning a motive spirit, as Aristotle says in his book *De spiritu* and elsewhere, is contractile, and in what way *neuron* is derived from *neuo*, that is I nod, I contract . . .

According to Harvey (whose etymology and definition are false), nerve is the generic term for everything that is contractile; and it comprises, as its species, ligament, tendon, origin of muscle, fibre and flesh (see *De motv locali animalivm 1627*, p. 79 *et passim*.).
⁶⁴ *Body*, HEW 1: p. 325.
⁶⁵ *Body*, HEW 1: p. 527.
⁶⁶ See Hahm, D. E. *The Origins of Stoic Cosmology*, Ohio State University Press (1977).

[67] For Hobbes' use of the terms 'cohesion', 'coherence' and 'incoherence', see *Body*, HEW 1: pp. 30, 57, 58, 215, 334, 388, 398, 399, 400, 445, 452, 476, 479; *Human Nature*, HEW 4: pp. 11, 14, 15, 25; *De corpore politico: or the Elements of Law*, HEW 4: pp. 155 *et passim; Leviathan*, pp. 90, 95, 602. Cognate terms include unity, agreement, concord, congruity, consistency, tenacity.

[68] Marcus Aurelius, *Meditations*, translated ... by M. Staniforth, Harmondsworth (1974), p. 170.

[69] In my opinion the chief text for Hobbes' development of these principles is Galilei, G., *Discorsi e dimostrazioni matematiche, intorno a due nuoue scienze* ..., Leyden (1638), which contains a theory of music. For the importance of this theory, see Cohen, H. F., *Quantifying Music: The Science of Music at the First Stage of the Scientific Revolution, 1580—1650*, Dordrecht (1984). The Stoics pictured tensional motion by analogy to water waves in a pool which expand in circles when a stone is thrown into the water. According to Cohen *ibid.*, this analogy was extremely fruitful for the development of the new science of musical acoustics. Shapiro *op. cit.* (n. 18) suggests that Hobbes' 1644 theory of light was modelled on musical acoustics.

[70] See n. 25 above.

[70] See n. 25 above.

[71] *Body*, HEW 1: pp. 93—202. This constitutes Hobbes' second section, 'The First Grounds of Philosophy', which, in my reading, is a continuation of section 1, 'Computation or Logic', since it deals with the categories of Stoic logic: body, accident, relation and relation-in-a-certain way. Indeed, Hobbes' odd statement about the Aristotelian categories ('predicaments'), *ibid.*, HEW 1: p. 28, makes sense only if this section is so understood.

[72] Vital motion: *Body*, HEW 1: pp. 406, 407; *Human Nature*, HEW 4: p. 31; *Leviathan*, pp. 118, 122. The vital motion of the blood is its external, circular movement, not its internal impulse.

[73] Interior beginnings of animal motion: *Body*, HEW 1: p. 407, *Human Nature*, HEW 4: pp. 31, 32; *Leviathan*, pp. 118—9. For animal motion in general: *Body*, HEW 1: pp. 405, 407, 408; *Human Nature*, HEW 4: pp. 30, 31, 32; *Leviathan*, p. 118.

[74] For a summary of Hobbes' epigenetic theory of knowledge, see 'The Answer of Mr. Hobbes to Sr Will. D'Avenant's Preface before Gondibert', in D'Avenant, W., *Gondibert: An Heroick Poem*, reprinted by Scolar Press (1970), p. 78. The first edition was published at Paris in 1650; another edition appeared in London in 1651.

[75] Hobbes defined habit in *Body*, HEW 1: pp. 348—50, where he also provided two examples, one of inanimate, the other of animate habit. In the inanimate example, he pointed out that a bow will become bent by the constant restraint of the string, an example that is reminiscent of the adage: 'as the twig is bent, so is the tree'. In the animate example, he had recourse to a person learning music. Hobbes was not the first, nor the last, person to illustrate the transition of voluntary actions into automatic ones by recourse to the way in which a musician learns to finger a musical instrument. For other such illustrations, see Kassler, J. C., 'Man — A Musical Instrument: Models of the Brain and Mental Functioning before the Computer', *History of Science* 24, 59—92 (1984).

[76] *I.e.*, ratiocination or computation (by adding and subtracting, multiplying and

dividing), *Body*, HEW 1: pp. 3—5. See also *ibid.*, HEW 1: pp. 408—9, where Hobbes defined deliberative processes as a succession of appetites and aversions (that is, conceptions of pleasure and pain), of which there are many degrees. The source of appetite and aversion is more or less heat, as Hobbes made clear in *Body*, HEW 1: p. 401 and, especially, in *Human Nature*, HEW 4: p. 8: '. . . our heat is *pleasure* or *pain*, according as it is *great* or *moderate*'.

[77] *I.e.*, self-conservation; see *De corpore politico: or the Elements of Law*, HEW 4: pp. 83, 85, 117.

[78] Deliberation is that time when 'the action is in our power to do or not to do', *Human Nature*, HEW 4: p. 69. Hence, it is a decision process, according to which the self will decide to give assent to or to withhold it from a conception. The last deliberation in this decision process constitutes the will, which is an act itself or a decision that an act is impossible. Thus, in *Body*, HEW 1: p. 409, Hobbes wrote: 'But if deliberation have gone before, then the last act of it, if it be appetite, is called *will*; if aversion, *unwillingness*.' If the decision is to act, then impulse (Harvey's 'nod', n. 63 above) functions as a stimulus to the muscles, enabling the appropriate local motion to take place.

[79] *Body*, HEW 1: p. 409; see also *Human Nature*, HEW 4: pp. 68—9.

[80] *Body*, HEW 1: pp. 55—64; *Human Nature*, HEW 4: pp. 23—4, 28—30.

[81] *Human Nature*, HEW 4: pp. 18, 29: 'Prudence is nothing but conjecture from experience, or taking of signs from experience warily, that is, that the experiments from which he taketh such signs be all remembered; for else the cases are not alike that seem so.' Accordingly, when '*experience of fact* is great, it is called prudence'. See also *De corpore politico: or the Elements of Law*, HEW 4: p. 110.

[82] *Body*, HEW 1: pp. 399, 409; *Leviathan*, pp. 93, 96, 127.

[83] *Leviathan*, p. 82; see also *Human Nature*, HEW 4: p. 26.

[84] For parts and wholes, see *Body*, HEW 1: pp. 9—10; for the unity of the self, see *ibid.*, HEW 1: p. 391: 'The *subject* of sense is the *sentient* itself, namely, some living creature; and we speak more correctly, when we say a living creature seeth, than when we say the eye seeth.'

[85] For sleep, dreams and sickness as symptoms of incoherence of the self, see *Body*, HEW 1: pp. 399—402. In *Human Nature*, HEW 4: p. 54, Hobbes made it clear that his focus was on healthy creatures, whose organs are 'well disposed', that is, 'equally tempered'. Hence, he used dreams rather than illness as the principal example of incoherence. See also *Human Nature*, HEW 4: pp. 10—1; *Leviathan*, pp. 90—3.

[86] *De corpore politico: or the Elements of Law*, HEW 4: p. 110: 'the habit of doing according to . . . [the] laws of nature, that tend to our preservation, is that we call *virtue*; and the habit of doing the contrary, *vice*. As for example, . . . temperance is the habit by which we abstain from all things that tend to our destruction, intemperance the contrary vice . . .' This is the Stoic version of *sophrosyne*, for which see the masterly study by North, H., *Sophrosyne: Self-knowledge and Self-restraint in Greek Literature* (Ithaca, 1966). The Stoic proverb, that intemperance is the mother of passion, was quoted by Cicero, *Tusculan disputations* 4.9.22, who also pointed out that temperance quiets the appetites and causes them to obey right reason; that is, temperance defends the judgments of reason against the assaults of passion.

[87] *Body*, HEW 1: pp. 135—8, 323—4. Accordingly, the principal differences between

humans are passions and strength. While both are powers, they are also limited powers and, so, not to be ascribed to the deity.

[88] Tracy, T. J., *Physiological Theory and the Doctrine of the Mean in Plato and Aristotle*, Chicago (1969), p. 358.

[89] I hope to develop Hobbes' conception of the self in a separate work.

[90] Hobbes assigned two causes of madness: one from the diseased constitution of the instruments of the body, the other from a vehement and too long continued passion. Thus, he included both our internal character and environmental considerations in his treatment of madness. See *Human Nature*, HEW 4: p. 54—9. Despite the different causes, Hobbes pointed out in *Leviathan*, p. 140, that 'in both cases the Madness is of one and the same nature', that is, contrary to reason.

[91] The collection was published posthumously in 1668. I have used Cowley, A., *The English Writings* edited by A. R. Waller, 2 Vols, Cambridge (1905—1906), 2: pp. 377—459.

[92] *Ibid.*, 2: p. 377.

[93] For a detailed study of Cowley's work, see Rostvig, M.-S., *The Happy Man: Studies in the Metamorphoses of a Classical Ideal*, 2d edn., 2 Vols, Oslo and New York (1962) 1st publd. (1954), who shows how the figure of the classical *beatus vir.*, re-interpreted in terms which suited the religious sensibility of the age, was definitely established during the Civil War as a Royalist counterpart to the Puritan pilgrim/warrior.

[94] Cowley, *op. cit.* (n. 91), 2: p. 384.

[95] *Ibid.*, 2: pp. 384—85. For the classical imagery related to *sophrosyne*, see North *op. cit.* (n. 86).

[96] Cowley *ibid.*, 2: p. 388.

[97] *Ibid.*, 2: p. 383; for the source, also of Hobbes' *Leviathan*, see the Book of Job 7:12, 41:1—34. *Pan huper sebastus: pan(h)upersebastos*, probably a nonceword and certainly Byzantine. It reflects the fact that the first teachers of Greek in England had come, as refugees from the Turks, from Constantinople (Anderson, W. D., personal communication).

[98] Cowley *ibid.*, 2: p. 459.

[99] *Ibid.*, 2: p. 384. See also Hobbes, *Leviathan*, p. 161: '... the voluntary actions, and inclinations of all men, tend, not only to the procuring, but also to the assuring of a contented life ...'

[100] Hobbes, *ibid.*, pp. 129—30.

[101] Cowley intended the *Discourses* as an explanation and defence of his retirement from public life after the Restoration. For evidence that his work embodies personal experience, see Rostvig, *op. cit.* (n. 93), 1: pp. 20—1, 35, 38, 212, *et passim*.

[102] Hobbes used the term 'light' and 'dark' (the privation of light) as metaphors for knowledge and ignorance. According to Barker and Goldstein, *op. cit.* (n. 6), p. 159, this tradition derives from Augustine of Hippo, who 'retained from the Stoicism of his pre-Christian period the metaphor of illumination as the causal mode connecting God and man'.

[103] In *Human Nature*, HEW 4: pp. 50—1, Hobbes indicated that our desire for knowledge is called 'curiosity' and from this arises 'not only the invention of names, but also supposition of such causes of all things as they thought might produce them. And

from this beginning is derived all *philosophy*'. In a letter, HEW 7: p. 467, prefixed to one of his unpublished optical treatises, Hobbes wrote:

The passions of man's mind, except onely one, may bee observed all in other living creatures. They have desires of all sorts, love, hatred, feare, hope, anger, pitie, aemulation, and ye like: onely of curiositie, which is ye desire to know ye causes of thinges, I never saw signe in any other living creature but in man.

In *Body*, HEW 1: p. 168, Hobbes hinted that curiosity (the desire to know causes) is the 'seed' of natural religion; see also *Leviathan*, pp. 124, 167, 172.

[104] *Body*, HEW 1: pp. 56—7.
[105] On this point, see Kieffer, J. S. (tr.), *Galen's Institutio Logica* ... , Baltimore (1964), pp. 1—18.
[106] Hobbes summarized his position in *Leviathan*, p. 106, as follows:

... between true Science, and erroneous Doctrines, Ignorance is in the middle. Natural sense and imagination, are not subject to absurdity. Nature it selfe cannot erre: and as men abound in copiousness of language; so they become more wise, or more mad than ordinary. Nor is it possible without Letters for any man to become either excellently wise, or (unless his memory be hurt by disease, or ill constitution of organs) excellently foolish. For words are wise mens counters, they do but reckon by them: but they are the mony of fooles, that value them by the authority of an *Aristotle*, a *Cicero*, or a *Thomas*, or any other Doctor whatsoever, if but a man.

In short, universal nature, not man, is the criterion.
[107] *Body*, HEW 1: pp. 55—6. Misnomers: calling 'any thing a name, which is not the name thereof'; hasty inference: 'pronouncing rashly'.
[108] *Ibid.*, HEW 1: p. 56.
[109] *Ibid.*, HEW 1: p. 57.
[110] *Ibid.*, HEW 1: pp. 87—8.

... such things as I have said are to be taught last, cannot be demonstrated, till such as are propounded to be first treated of, be fully understood. Of which method no other example can be given, but that treatise of the elements of philosophy, which I shall begin in the next chapter, and continue to the end of the work.

The term 'work', of course, could refer not only to *Body* but also to *Human Nature* and *De corpore politico: or the Elements of Law*.
[111] Pre-rational, *i.e.*, before 'wit' (reason) has been acquired 'by method and instruction'; see Hobbes, *Leviathan*, pp. 134—9. Accordingly, pre-rational includes children and fools (*i.e.*, dullards).
[112] For the digressions, see *De corpore politico: or the Elements of Law*, pp. 120—21; *Leviathan*, pp. 225—27. The notion that 'the Ant is ... the Emblem of Self-wisdome' was widespread; see Lansdowne, the Marquis of (ed.), *The Petty-Southwell Correspondence 1676—1687* ... , London (1928), p. 289.
[113] *Leviathan*, pp. 404, 534, 576.
[114] For the classical antecedents, see Anderson, W. D., *Ethos and Education in Greek Music*, Cambridge, Mass. (1966).

[115] Hobbes defined study in *Body*, HEW 1: p. 395; but his method of study is only now coming to be understood, owing particularly to the careful readings of William Sacksteder. See, especially, Sacksteder, W., 'Hobbes' *Logistica*: Definition and Commentary', *Philosophy Research Archives* **8**, pp. 55—94, (1982).

[116] *I.e.*, philosophy or knowledge of causes and effects; see *Body*, HEW 1: p. 3; *Human Nature*, HEW 4: pp. 28—9.

[117] *Leviathan*, pp. 216—7. According to Hobbes (*ibid.* p. 545), the precepts of natural reason are 'written' in the heart.

[118] In *Leviathan*, pp. 454—5 *et passim.*, Hobbes abbreviated the two laws to 'reason' (universal nature/*logos*, *i.e.*, sympathetic resonance) and 'equity' (peace/love, *i.e.*, concord).

UDO THIEL

CUDWORTH AND SEVENTEENTH-CENTURY THEORIES OF CONSCIOUSNESS

One of the few things that most philosophers working within various traditions agree about is that the notion of consciousness is a key, or fundamental notion, both in epistemology and in any theory of the self. Discussions of consciousness as we know them began in earnest in the late seventeenth/early eighteenth century — largely under the influence of Cartesianism. This at least is the acknowledged historical source of present-day contributions to the theory of consciousness. In this paper I am concerned with another treatment of the notion of consciousness in early modern philosophy, the importance of which has not been sufficiently recognized by scholars, namely the treatment of the issue by the Cambridge Platonist Ralph Cudworth (1617—1688); and I shall examine and evaluate Cudworth's contribution by way of considering it in the context of seventeenth century thought.[1] Now, some might argue that it is rather misleading to speak, as I do in the title of this paper, of seventeenth century *theories* of consciousness, for, so it might be said, even though Cartesians and some other philosophers may have raised a number of issues which are relevant to a theory of consciousness, there really were no worked out theories of consciousness in the seventeenth century. However, although it may be true that in the seventeenth century there were no elaborate theories of consciousness, no monographs, no chapters of books devoted to the topic, there were certainly various attempts in the context of debates on other issues to become clear about the notion of consciousness and related concepts. And one very important attempt is, I argue, to be found in the works of Ralph Cudworth.

The English terms 'consciousness' and 'being conscious' were not used frequently at all until late in the seventeenth century. Indeed, Cudworth was the first English writing philosopher to make extensive use of the noun 'consciousness' and to attach to it a particular, definitely philosophical meaning. In my view, Ralph Cudworth should be credited with having introduced 'consciousness' as a philosophical term into English (even though other English writers had occasionally used that term before him, for example his fellow Platonist Henry

More), just as Christian Wolff is credited with having introduced the German counterpart 'Bewusstsein' into German philosophical terminology.[2] Whereas Wolff introduced the German term as a translation of the Latin and French terms which he found in philosophers like Descartes and Leibniz, Cudworth made explicit recourse to Neo-Platonic sources, and especially to Plotinus. Thus, at least in England, the term 'consciousness' came onto the scene not primarily as a Cartesian influence (although that influence is also relevant in Cudworth), but much more as part of a revival of the Platonic tradition. This points to the particular context in which the concept of consciousness is developed in Cudworth. I said that the term 'consciousness' was not used frequently at all until late in the seventeenth century. By the turn of the century the term had become an established one in philosophical debates; but even then the notion of consciousness itself was not made an explicit object of philosophical enquiry; the term was used in various senses, and it is not clear what exactly the term 'consciousness' was thought to denote. It was not until the late 1720s that more substantial discussions of consciousness appeared: Chamber's *Cyclopaedia* of 1728 has a separate entry for 'consciousness' (which is not true of John Harris' *Lexicon Technicum* of 1710 for example); and the notion of consciousness plays an important role in Peter Browne's *The Procedure, Extent, and Limits of Human Understanding* (London, 1728); more importantly still, in the same year, 1728, an essay solely devoted to the notion of consciousness appeared anonymously in London.[3] I would like to cite another work from that period which will lead us straight back to the seventeenth century and Ralph Cudworth: In 1727 John Maxwell published as an appendix to his translation of Richard Cumberland's *De Legibus Naturae* a (very useful) summary of the controversy between Samuel Clarke and Anthony Collins concerning the immateriality and natural immortality of the human soul.[4] In this controversy extensive use was made of the term 'consciousness'.[5] Maxwell had realized, though, that it was being used in various senses, and he regarded it as appropriate to preface his summary by a note on that very term. He points out that 'in the following Reasoning' the term 'consciousness' can mean (a) 'the reflex Act, by which a Man knows his Thoughts to be his own Thoughts', (b) 'the Direct Act of Thinking; or (which is of the same Import;) simple Sensation', or even (c) 'the Power of Self-motion, or of beginning of Motion by the Will'. Maxwell believes that the argument for immateriality holds whichever notion of consciousness we apply; nevertheless, he does not fail to point out that the first, i.e. consciousness as a 'reflex act' on our own thoughts, is 'the strict and properest Sense of the Word'.[6] Consciousness, then, is understood here as a way of referring to one's own self. And, indeed, it is in this sense of an individual self-reference that the concept of consciousness had developed in the seventeenth century. Yet, to say that consciousness was thought of as a form of individual self-reference is still rather vague; we need to determine *what* form of self-reference consciousness was held to be, and how it was thought to relate to other forms of self-reference, such as *self*-consciousness, self-knowledge, introspection, reflection and even conscience. I shall argue that, even though Cudworth, too, does not have a worked out theory of consciousness, there are conceptual distinctions essential to such a theory present in his writings, and that Cudworth's distinctions are made more explicitly and more clearly than in other seventeenth century authors before him. So, in the remainder of this paper, I shall do two things: first, I offer my analysis of seventeenth century discussions prior to Cudworth and the Cambridge Platonists; this, as presented here, can only be very brief and sketchy; it is impossible to cover all of the often very complex material in one paper. Second, I seek to analyse and evaluate Cudworth's notion of consciousness in the light of these debates.

I said that Cudworth made explicit recourse to ancient sources when employing the notion of consciousness. And, indeed, the problem of self-reference is an old one. There are well known passages in Plato's Dialogue *Charmides*, for example, in which the problem of reflective knowledge is examined in the context of an analysis of the concept of *sophrosyne* (temperance).[7] And there are passages in the various works of Aristotle, especially in *De Anima*, where the problem of self-reference is addressed.[8] Maxwell's definition of consciousness as a 'reflex act', however, provides a guide to which notions in particular we must look when examining seventeenth century conceptions of consciousness, namely to the notions of *conscientia* and *reflexio*. Like 'conscience', and the French 'conscience' of course, 'consciousness' stems from the Latin 'conscientia', the noun for 'con-sc/i/re'. And 'conscientia' is in turn a translation of Greek terms, the main candidates being 'syneidesis', 'synesis', and, especially relevant for Cudworth, 'synaisthesis'.[9] Initially, the Greek 'syneidesis', the Latin 'conscientia', and even the English 'consciousness' or 'being conscious' were not self-referential at all, but meant a perception or knowledge of something that one shares with someone else: being conscious meant being privy to something. In fact, this is how 'being conscious' was still defined occasionally in the seventeenth century.[10] However, this was not the meaning that was adopted by the Cambridge Platonists and other philosophers in the second half of the seventeenth century. Like the Greek 'syneidesis' and the Latin 'conscientia' (much earlier), 'consciousness' changed its meaning from 'knowing together with someone else' to 'knowing something for oneself': the person with whom I am privy to something is not someone else. Thus, 'conscientia' (and 'consciousness') came to be used as referring to a knowledge (however that may be defined) of one's *own* states and acts, that is, it became truly self-referential. The English 'conscience' is much older than 'consciousness', and it was very common indeed throughout the seventeenth century and even before that time. It derives from a further development of 'conscientia' denoting a moral judgement of one's own actions and thoughts. Again, 'syneidesis' and 'conscientia' had adopted this special meaning of the judgement or knowledge of one's own right and wrong doing much earlier. In the seventeenth century 'conscientia' covered both conscience and the non-evaluative knowledge of one's own actions. Similarly, the English 'conscience', although it has moral overtones, partly covered that meaning which later came to be expressed independently by the term 'consciousness'. The other crucial term, 'reflexio' or 'reflection', can also mean various things. In its most general sense it simply means discursive thinking or considering. I am concerned with its narrower and more interesting sense as a 'return to oneself'. This theme of turning one's attention away from the external world to one's own self is much older than the philosophical term 'reflection'; it has been discussed ever since antiquity. An often quoted author in the seventeenth century is Augustine who suggests that the way to discover truth is the 'reditio in se ipsum', the 'return to oneself'.[11] The point is taken up by Aquinas who argues that the intellect knows truth because it reflects on itself, by which he means that the intellect not only knows that it knows, but also knows its own essence.[12] We shall see that the theme of a 'return to oneself' is also present in Cudworth's account of knowledge.

Now, prior to the Cambridge Platonists, there were two major seventeenth century thinkers or groups of thinkers who dealt to a considerable extent with issues concerning self-reference: First, there were the scholastics; the textbooks and dictionaries of seventeenth century 'schoolmen' like Goclenius and Micraelius, for example, were a powerful force throughout the period. Second, and more obvious, there were the Cartesians whom I mentioned earlier. The notions of *reflexio* and *conscientia* were part of scholastic doctrine and teaching both before and during the seventeenth century. It was standard practice in the textbooks to distinguish thought, understanding and knowledge into *direct* and *reflex* thought, understanding, and knowledge. And *reflexio* in general is said to be an explicit 'return' of the mind or intellect to itself. Both Goclenius and Micraelius tell us that that act of the mind is reflexive whereby the mind understands itself.[13] The account given of *reflexio* is, in general, pretty much the same as what one can draw together from various remarks Aquinas had made on the topic. For Aquinas, reflection relates to mental acts and their contents as its objects; this means, second, that reflection is to be understood as an act distinct from those acts it reflects upon — as Aquinas says: '... the act whereby the intellect understands stone is different from the act whereby it understands that it understands stone';[14] third, since, according to Aquinas, 'more than one thing cannot be understood at a time',[15] it follows that the act of reflection can take place only *after* the act reflected upon has occurred; in other words, reflection relates necessarily to one's past only, it can't refer to present acts of the mind. These characteristics of reflection re-appear in the seventeenth century textbooks; Goclenius, for instance, says that the understanding reflects when, after having understood other things, it returns to itself and understands its previous acts of understanding.[16]

As far as I know, *reflexio* and *conscientia* were never discussed together or explicitly related to one another, probably because, unlike *reflexio*, *conscientia* formed a very special topic within moral theology; nevertheless, as we shall see, the two concepts are related. I should say that *conscientia* was not, of course, a topic only for the scholastics. In England, for example, there were countless sermons and tracts on the problem of *conscientia*, especially in the context of puritan teaching and its emphasis on the individual's conscience. The picture one gets, however, is that even here the general notion of *conscientia* that is applied is very much the same as in traditional scholastic thought. *Conscientia* was thought to consist of three 'parts' or elements: the first is a set of objective moral principles which set the standard according to which we ought to direct our actions; this is sometimes referred to as the rational or 'pure' part of conscience; the second is a knowledge or remembrance of thoughts and actions performed by us. Anthony Cade described it in 1621 as 'a Chronicle, or register, roll or record' where all our 'thoughts, words, and actions be they good or evil' are set down. The third part is the moral *judgement* of our remembered actions on the basis of the moral principles.[17] One can easily see that in this account the older, non-evaluative sense of 'conscientia' is retained as the second part or element, i.e. as the remembrance of our thoughts and actions: the non-evaluative knowledge or remembrance of our own actions is a prerequisite for our moral judgement of those actions. And there is an apparent similarity between this non-evaluative *conscientia* and *reflexio*: not only are both, quite obviously, self-referential, but they are both acts distinct from the ones they relate to, and they relate to *past* thoughts and actions only. On the other hand, *reflexio* differs from *conscientia* as 'chronicle' at least in this, that it does not only relate to thoughts and actions that are morally relevant, but to all thoughts and mental operations whatsoever. *Conscientia* as 'chronicle' is just an element of the moral conscientia-theorem as a whole and not to be dissociated from it. Sometimes, the term 'conscientia' was used to denote the traditional notion of *sensus communis*. This is true, for example, of Herbert of Cherbury who defines *conscientia* as the '*sensus communis* of the inner senses'. However, here too, the function ascribed to *conscientia* is mainly a moral one.[18]

The fact that reference to one's own mental operations plays a central and systematic role in Descartes' philosophy is well known. Descartes and his followers tried to re-establish a notion of non-evaluative *conscientia* that is independent of moral theology and to attach to it a central role in philosophical enquiry.[19] And although even the Cartesians did not develop a systematic *theory* of consciousness as a particular form of self-reference, there are certainly elements for such a theory in the writings of some of them.[20] Quite unlike the scholastics, Descartes defines thought in terms of consciousness by saying that thought includes everything that is in us in such a way that we are immediately conscious of it.[21] The notion of 'conscium esse' or 'conscientia' that is implied here seems to be different from both the traditional concept of non-evaluative *conscientia* and of *reflexio*. For Descartes, consciousness seems to be an essential element of thought itself; 'conscientia' seems to denote a reflexivity inherent in all thought as such. In a letter to Arnauld he makes a point of distinguishing between 'being conscious' and 'remembering', which means that for him non-evaluative *conscientia* is *not* a 'chronicle or register'. Furthermore, in the same letter, he draws a distinction between direct and reflex thought, implying, so it seems, that his 'conscientia' is *not reflexio*.[22] Yet, the problem is that when he actually defines consciousness, he defines it in terms of reflection. He says that 'to be conscious is both to think and to reflect upon one's thought'.[23] Thus, while Descartes dissociates consciousness from moral theology, his concept of *conscientia* appears to be permeated by the notion of reflection. It differs from the scholastic notion of reflection in one important respect, though, namely in that it indicates a reference to the present. Against Burman he argues that 'it is false that this reflection [i.e. consciousness] cannot occur while the previous thought is still there. This is because ... , the soul is capable of thinking of more than one thing at the same time, and of continuing with a particular thought which it has'.[24] Unfortunately, Descartes does not elaborate any further on the concept of consciousness and related concepts. There are passages where he speaks of 'reflective knowledge' or a knowledge of the nature of thought 'acquired by demonstration' as being always preceded by 'internal cognition'. However, one cannot ascribe to Descartes a clear distinction between consciousness and reflection on the basis of such passages.[25] For this 'reflective knowledge' is more like what one could call 'philosophical reflection'; and I do not wish to suggest that Descartes was not aware of the difference between *that* and *conscientia*: there is an obvious distinction implicit in Descartes between *conscientia* and philosophical reflection *about conscientia* in general, thought in general, mind in general and so on.[26] But, of course, it is not *this* distinction that is at issue here. What is at issue are the various forms of an individual's relation to her- or himself, and on this Descartes does not provide a theory, nor does he provide the conceptual distinctions that are relevant for such a theory.

The notion of consciousness as self-reference can be found in many writings of Descartes' followers and critics in the 17th century. Hobbes, for example, argues in the *Leviathan* 'that being awake, I know I dreame not; though when I dreame, I think my self awake'.[27] This seems to suggest that he accepts the idea of a reference to one's own present mental states. Yet, elsewhere he explicitly denies the possibility of self-reference to present thoughts. In his second objection to Descartes he argues against the latter's notion of *conscientia* saying that 'though someone can think he *has* thought ... nevertheless it is quite impossible to think that one is thinking, and to know that one is knowing. This question would go on ad infinitum: How do you know that you know you are knowing?'[28] Hobbes obviously assumes that Descartes' *conscientia* is a separate act directed to one's own present thoughts, and he argues that the only reasonable notion of *conscientia* is in terms of scholastic *reflexio* or remembrance.

Despite this critique, the idea that non-evaluative conscientia accompanies our present thoughts was adopted by most thinkers who worked in the Cartesian tradition throughout the second half of the seventeenth century. Philosophers like Louis de la Forge, Cordemoy, Geulincx, Clauberg and later, in Germany, Tschirnhaus and Christian Thomasius all asserted that consciousness relates to our present thoughts. Louis de la Forge is important in that he introduced the French term 'conscience' in its non-moral sense into the philosophical debate.[29] Yet, the notion of consciousness itself is hardly ever explained in its own right by these authors. The concepts of reflection, conscience, self-knowledge and so on are not really sorted out from one another. A more careful discussion of some of these concepts in the broad Cartesian context can be found later in the century in Antoine Arnauld, especially in his critique of Malebranche. However, this critique was published five years after Cudworth's main work, and is therefore not relevant in the present context.[30] Malebranche's *De la Recherche de la Vérité* was published in 1674, that is four years before Cudworth's *True Intellectual System of the Universe* appeared, but three years after that huge work had been given its *imprimatur*. Malebranche had challenged the Cartesian assumption that *conscientia* provides the basis of complete and absolutely certain knowledge of the human soul. According to Malebranche, we can attain knowledge proper of external objects through ideas that represent them; these ideas are real beings in the divine mind. But our own souls, says Malebranche, we don't know through ideas in God; we know our own souls only through a 'sentiment interieur' for which he — like de la Forge before him — also uses the term 'conscience'.[31] And it is precisely because our knowledge via *conscience* is not mediated through God, that it must be imperfect knowledge; since *conscience* is the only basis for knowledge of the soul, we cannot arrive at 'une entiere connaissance de notre ame'.[32] Even though the notion of *conscience* is obviously crucial to Malebranche's theory, he, too, does not elaborate on it. It remains unclear exactly what type of 'knowledge' or self-reference *conscience* is. This is true also of other thinkers of the time. Mostly, it seems, it was simply assumed that self-reference via non-evaluative *conscientia* is to be equated with an act of reflection which requires special attention or with a function of the inner sense. Matthew Hale, for example, writes that having knowledge of one's own thinking is a 'reflex act of the Soul, or the turning of the intellectual eye inward upon its own actions'.[33] And John Wilkins states that the 'inward sense' is that 'by which we can discern *internal* objects, and are conscious to our selves, or sensible both of the impressions that are made upon our outward *senses*, and of the inward motions of our *minds*'.[34] Many authors, not only the scholastics, *applied* the notion of reflection, for instance in the context of an argument for the immateriality of the soul,[35] without, however, entering into a discussion of what reflection is and how it may be distinguished from other forms of relating to oneself. It is only, as far as I can see, in the writings of Ralph Cudworth that a distinction can be found between consciousness as the most fundamental form of self-reference and other types of reference to the self.

Now, even though Cudworth makes extensive use of the term 'consciousness' and draws certain important distinctions, consciousness does not form a separate, special topic of discussion for him; there is no special chapter entitled 'on consciousness'. The reader needs to extract Cudworth's view from various works, including, especially, various passages from the 900 folio pages of *The True Intellectual System of the Universe* (London 1678; abbr.: TIS), and also from the posthumously published *A Treatise concerning Eternal and Immutable Morality* (London 1731; abbr.: EIM) and *A Treatise of Freewill* (London 1838; abbr.: TFW). Furthermore, it is important to bear in mind that Cudworth's statements about consciousness are, mostly, not part of an analysis of human subjectivity, but of a metaphysical account of reality in general — an account which affirms the traditional idea of a scale of nature, drawing heavily on Plotinus (TIS 648). Indeed, Cudworth first raises the problem of consciousness in the context of a discussion of causation in nature (TIS 146—181). He argues against both a purely mechanistic explanation of events in nature and a theological voluntarism, according to which God always directly intervenes as the cause of events in nature. According to Cudworth, it is neither reasonable to assume that everything happens through a blind mechanism of matter, nor is it reasonable to assume that God does everything himself directly and via miracles (TIS 147). In addition to mechanism and the divine will, Cudworth argues, we have to assume a third power which is a divine instrument, producing effects by executing the divine will. This power is what Cudworth terms the 'Plastic Nature of the Universe'. The plastic nature is responsible for the ordering of matter in the forming of plants, animals, and other corporeal beings. Cudworth even says that the laws of nature are nothing else but a plastic nature which works in the whole corporeal universe (TIS 151). As a divine power it must be thought of as being immaterial, but as God's lowest power it is capable of uniting itself with matter and of working on matter. The plastic nature is, as Cudworth says '... *Reason Immersed and Plunged* into Matter, and as it were *Fuddled* in it, and *Confounded* with it. *Nature* is not the *Divine Art Archetypal*, but only *Ectypal*' (TIS 155, par. 11). Cudworth does not, of course, claim that this notion of a plastic nature is original to him. Other Cambridge thinkers like Henry More adopted it (he calls it the 'Spirit of Nature'); and Cudworth himself, as always, goes to great length and detail in attempting to show that the notion can be found in ancient Greek thinkers (TIS 151—4). It is certainly present in Stoic thought as 'logoi spermatikoi', 'seminal reasons' or 'forming forces'; and it corresponds to the lower level of the World-Soul in Plotinus (*physis*).[36]

What has all this to do with consciousness? Well, Cudworth points out that since the plastic nature is the lowest of the divine powers, it has certain deficiencies; and these deficiencies relate to self-reference. First, although the plastic nature acts *according* to divine wisdom and fulfils divine purposes, it does not itself apprehend and understand those purposes and the reasoning behind what it does. It has no knowledge of the purposes and reasons of those acts it performs. Nature is, as Cudworth says, 'a living Stamp or Signature of the Divine Wisdom, which though it acts exactly according to its *Archetype*, yet it doth not at all Comprehend nor Understand the Reason of what it self doth' (TIS 155, par. 11). It is 'not *Master* of that *Consummate Art* and Wisdom according to which it acts, but only a *Servant to it*, and a *Drudging Executioner* of the Dictates of it' (TIS 156, par. 12). Second, apart from this lack of knowledge of the reasons and purposes for its actions, it has another deficiency: it does not even know that it performs the actions it does perform. So, Cudworth distinguishes two forms of self-reference here: (1) knowledge of the reasons and purposes for one's own actions, and (2) knowledge of the fact that one performs the actions one does perform. It is to describe this second kind of 'knowledge' that Cudworth uses the terminology of consciousness. Cudworth says that the plastic nature is not '*Clearly and Expresly Conscious of what* it doth' (TIS 158). And consciousness is understood by Cudworth as an immediate feeling or perception of one's own thoughts and actions while one is performing them. For it is clear from the text that Cudworth uses 'consciousness' as a translation of the Greek 'synaisthesis' which he found in Plotinus. He was obviously looking for a suitable English term for this special self-reference to one's own actions. He says that the plastic nature has 'no Express *synaisthesis, Con-sense* or *Consciousness* of what it doth' (TIS 159).[37]

According to Cudworth, the plastic nature is incorporeal, but without consciousness; and this is clearly a rejection of a Cartesian-type dualism, according to which there are, apart from God, only two kinds of sharply distinguished substances, *res cogitans* and *res extensa*. Cudworth explicitly states that it is not extension and thought which are the 'first Heads of Being', but extension and *life* (TIS 159). Unlike extension, life has an internal power and self-activity, and is essentially immaterial. Plastic powers belong to life and are, therefore, immaterial. Yet, as we have just seen, not all actions which emerge from the immaterial life are accompanied by consciousness. In Cudworth, the distinction between materiality and immateriality is not co-extensive with that between thought and extension.[38] This becomes clearer when we consider Cudworth's notion of individual plastic powers; there is not only a general plastic nature of the universe, but there are also unconscious plastic natures at work in each individual living being (TIS 167/171). For Cudworth, human souls are, just as for Descartes, immaterial. However, it is by making use of the very concept of consciousness that Cudworth distinguishes between two kinds of incorporeal mental life: he distinguishes between a pure rational part of the soul which is conscious, and an unconscious power of the soul which is responsible for organic functions, reflex actions, instincts, habits, and dreams. Cudworth terms this power the soul's 'plastic nature' (TIS 160–1). Thus, although Cudworth says that 'Consciousness . . . (is) essential to Cogitation' (TIS 871), he can still argue against Descartes' view according to which conscious activity is essential to the *soul*. On the Cartesian position, Cudworth says, one could not explain how the soul during sound sleep and fainting fits is without consciousness and still exists; it is obvious that the soul does exist in those states. Cudworth concludes: 'Now if the Souls of Men and Animals be at any time without *Consciousness* and *Self-perception*, then it must needs be granted, that Clear and Express *Consciousness* is not Essential to *Life*' (TIS 160).

Cudworth's account of the relationship of consciousness to thought, sense-perception and knowledge is to be found in the epistemological considerations of his *Treatise concerning Eternal and Immutable Morality*. Here, Cudworth argues that knowledge in general is itself to be understood as a form of self-reference.[39] He takes up and makes explicit the old idea that knowledge proper can only be attained by a 'return to oneself': 'The Essence of nothing is reached unto by the Senses looking Outward, but by the Mind's looking inward into it self. That which wholly looks abroad outward upon its Object, is not one with that which it perceives, but is at a distance from it, and therefore cannot Know and Comprehend it; but Knowledge and Intellection doth not not meerly look out upon a thing at distance, but makes an Inward Reflection upon the thing it knows, and according to the Etymon of the Word, [Intellectus] *the Intellect* doth [in Interioribus legere] *read inward Characters written within it self*, and Intellectually comprehend its Object within it self, and is the same with it' (EIM 97—8). Thus, knowledge in general is defined in terms of self-reference as 'inward reflection'; *this* reference to the self must be distinguished not only from consciousness and from knowledge of one's own operations, but also from self-knowledge understood as knowledge *of* the *self*. This distinction is not drawn explicitly by Cudworth, but it is clearly implied in his account.

Furthermore, Cudworth distinguishes in good Platonic tradition between 'sense' and 'knowledge'. To explain this distinction he first tells us what 'sense' is (EIM 75—84). And he argues against what he takes to be Hobbes' mechanistic view of sensation, according to which sensation is a mere corporeal, passive reaction to outward impulses.[40] For Cudworth, corporeal impulses become sensations only if we become *conscious* of these impulses as impressions on us. And since sensation requires consciousness, there must be an incorporeal mental part involved in sense-perception: To Cudworth, sensation arises when the soul, due to its union with the body, becomes *conscious* of those impulses. As he says: '. . . *Sense* is . . . a Cogitation, Recognition or Vital Perception and Consciousness of these Motions or Passions of the Body, therefore there must of necessity be another kind of Passion also in the Soul or Principle of Life, which is vitally united to the Body, to make up Sensation' (EIM 78).[41] Furthermore, this account of sensation is directed not only against Hobbes, but also against Descartes, because to say that sensation requires consciousness is to say that all sentient beings have consciousness (*synaisthesis*) and, therefore, have an immaterial soul — obviously an anti-Cartesian idea. Cudworth's *cogitations* include both sense-perceptions ('sensitive cogitations') and what he calls 'pure cogitations'; these are concepts which have their origin in the understanding itself, as for example concepts of relations (EIM 81).[42] According to Cudworth, only if we apply these pure concepts to what is given by sensation can knowledge of an object as a whole be attained: 'For Sense only takes Notice of several Colours and Figures either in the outside or the inside of any Animals, but doth not sum them up into one *Whole*' (EIM 170). It seems obvious for Cudworth that the soul has a consciousness of those pure concepts, since they are its own products; Cudworth says that 'the Mind is Naturally conscious of its own Active Fecundity' (EIM 138). The soul is conscious of its own products; it knows that the pure concepts have their origin in the soul itself and are *cogitations* distinct from sense-perceptions. In relation to what he calls '*Phantasms* or Imaginations' (EIM 138f.), Cudworth says: 'Now in this Case, when the Soul is conscious to it self, that these Phantasms are Arbitrarily raised by it, or by its own Activity, it cannot look upon them as Sensations' (EIM 118).[43]

We saw that consciousness is, for Cudworth, a feeling or perception of one's own sensations, thoughts and other activities; it is essential only to *cogitations*, the acts of the plastic nature are not accompanied by consciousness. Cudworth points out the importance of this self-reference via consciousness; it is the most fundamental relation of oneself to oneself: consciousness is that 'which makes a Being to be Present with it self' (TIS 159). Cudworth qualifies consciousness almost always by adjectives like 'clear' and 'express' (e.g. TIS 159—160). He thereby indicates that it is an explicit reference to the self where the self is the subject (i.e. that which is conscious) as well as the object of consciousness. This is what Cudworth has in mind when he says that a '*Duplication* . . . is included in the Nature of *synaisthesis, Con-Sense* and *Consciousness*' (TIS 159). It is this presence to oneself through consciousness which makes states of happiness and misery possible: Cudworth holds that this 'Duplication' via consciousness makes a being 'to perceive it self to Do or Suffer, and to have a *Fruition* or *Enjoyment* of it self' (ibid.). More importantly, consciousness as the most fundamental relation of oneself is that which makes other forms of self-reference possible. These other forms of self-reference are:

Self-consciousness. By 'self-consciousness' Cudworth means an assurance of the existence of one's own soul. And although 'that *Duplication* that is included in the nature of *synaisthesis*' is certainly relevant here, consciousness as *synaisthesis* is not itself self-consciousness; self-consciousness is not immediately given with *synaisthesis*. Cudworth says: 'we are certain of the Existence of our own Souls, partly from an inward *Consciousness* of our own *Cogitations*, and partly from that *Principle of Reason*, That, *Nothing can not Act*' (TIS 637). In other words, the certainty of one's mental existence is *inferred* from the fact that one can ascribe certain actions to oneself on the basis of *synaisthesis*. This means that Cudworth does not have to commit himself to the view that all sentient beings (who have consciousness) have self-consciousness as well; it means that only rational beings can have self-consciousness.

In the case of rational beings Cudworth distinguishes further forms of self-reference, in addition to *synaisthesis* and self-consciousness: We saw when discussing the notion of a universal plastic nature that Cudworth draws a distinction between consciousness of actions and the knowledge of the reasons and purposes behind those actions. In regard to the human soul Cudworth points out a corresponding difference between self-reference as consciousness and as *reflection*. Reflection, Cudworth suggests, occurs after we have performed the action we reflect upon (this is the old traditional notion of *reflexio*); and it can relate not only to actions we were conscious of at the time we performed them, but also to 'our doing ... *Animal* Actions' which we perform '*Non-attendingly*' (TIS 160). This reference to the self through reflection is contemplating and thus objectifying it. It is, like consciousness, a *duplication* of the self, but on a higher, rational level. As Cudworth says in the *Treatise on Freewill*: 'We are certain by inward sense that we can reflect upon ourselves and consider ourselves, which is a reduplication of life in a higher degree; for all cogitative beings as such are self-conscious' (TFW 71).[44] Whereas consciousness is essential to all *cogitations* and can thus be ascribed to all sensitive beings, i.e. to all beings that have (at least 'sensitive') *cogitations*, reflection is a different form of self-reference which is possible only on a higher, rational level of life. We saw above that Cudworth defines knowledge proper in terms of 'inward reflection'; clearly, his notion of reflection implies that self-reference which constitutes knowledge is not available to all sensitive beings, but only to those who have rationality and reflection as well.

Further, this self-reflection is the basis of, and can result in a self-*judgement*. And it is for this further self-reference, in the sense of the judgement of one's own self, that Cudworth uses 'conscience'. He obviously realizes that this term was also used to refer to what we mean by 'consciousness'; for he speaks of 'conscience in a peculiar sense' (TFW 71) when he means this self-judgement. We can see how the etymologically related terms 'consciousness' and 'conscience' are conceptually related in Cudworth: Self-reference through consciousness and self-reference through conscience belong to two different levels of life. However, beings that have conscience also have consciousness which, as the most basic kind of self-reference, is presupposed by conscience. As Cudworth says, it is through consciousness that we become 'Attentive to . . . (our) own Actions, or Animadversive of them' (TIS 159). For conscience to be possible, the capacity to reflect and a knowledge about the distinction between right and wrong must be presupposed in addition to mere consciousness: Conscience, Cudworth argues, is 'attributed to rational beings only, and such as are sensible of the *discrimen honestorum* or *turpium* when they judge of their own actions according to that rule, and either condemn or acquit themselves' (TFW 71).

According to Cudworth, self-reference via consciousness and reflection points at least to the possibility of another way of referring to oneself, namely to the possibility of self-determination with respect to action. He addresses the determinists' argument that self-determination is conceptually impossible, because one and the same being cannot be the subject (i.e. that which is determining) as well as the object of the determination (TFW 69). Cudworth argues that we have no reason to believe that such a relation of oneself to oneself is impossible; for there is an analogous 'duplication' or relation to oneself involved in consciousness and reflection. And a being that relates to itself in those ways should also be capable of being active on itself by determining its own actions: 'Wherefore that which is thus conscious of itself, and reflexive upon itself, may also well act upon itself, either as fortuitously determining its own activity, or else as intending and exerting itself more or less in order to the promoting of its own good' (TFW 71).

So far I have referred to Cudworth's *consciousness* or *synaisthesis* as the most fundamental relation of oneself to oneself. However, Cudworth indicates — by way of appealing to Plotinus — that there may be another, even more fundamental form of referring to oneself. We saw that, according to Cudworth, the plastic nature has no 'clear and express' consciousness of what it does; it is not present to itself. But there is one passage where he cites 'the often Commended Philosopher' (TIS 159) as conceding that nature has some basic sense of its own actions. He quotes from *Ennead* III, 8, 4 and translates: '*If any will needs attribute some kind of Apprehension* [synesis] *or Sense* [aisthesis] *to Nature, then it must not be such a Sense or Apprehension, as is in Animals, but something that differs as much from it, as the Sense or Cogitation of one in a profound sleep, differs from that of one who is awake*' (TIS 160). And Cudworth adds that 'it cannot be denied but that the *Plastick Nature* hath a certain *Dull* and *Obscure Idea* of that which it Stamps and Prints upon Matter' (ibid.). The sentence Cudworth quotes does not contain his favourite term for consciousness, 'synaisthesis'; but the term is used by Plotinus in the same section where he ascribes to nature 'a kind of self-perception' ('hoion synaisthesis'). Nature, then, has in Plotinus and Cudworth, a basic awareness of itself, a dim or 'dull' feeling of itself which does not involve a 'duplication', a division of a being into subject and object, as 'clear and express' consciousness does. This basic awareness, we are told by Plotinus-Cudworth, can be compared to that of a sleeping man who is in peace with himself.[45] Yet, Cudworth does not elaborate on this conception of a more basic form of consciousness which does not involve a 'duplication'. Cudworth is content to quote Plotinus as an authority on this. There is no clear distinction in Cudworth between consciousness as *synaisthesis* and as a 'certain *Dull* and *Obscure Idea*' of the self.

However, apart from this unclarity, there are, as we have seen, sophisticated conceptual distinctions concerning self-reference present in Cudworth. We can sum up Cudworth's views on self-reference as follows. Consciousness as *synaisthesis* is the fundamental type of self-relation. It is an intimate feeling of one's own *cogitations* and can only be ascribed to incorporeal life; yet it is not essential to life, for there is life which acts unconsciously (plastic natures). Consciousness is relevant to our assurance of the existence of our souls as the bearers of those *cogitations*. Yet, this assurance requires reasoning and is therefore only attributable to rational beings. This is also true of self-reference via the objectifying reflection and other, 'higher degrees' of self-reference such as the practical self-relation in conscience and in self-determination. No one before Cudworth seems to have drawn these distinctions as explicitly and clearly as they are present in his writings. And although Cudworth, like the other Cambridge Platonists, was and is regarded as an 'unmodern' thinker, we know that his theory had a considerable impact on subsequent thought. It has been shown, for example, that Locke developed his own new and immensely influential theory of personality in terms of consciousness precisely at the time when he was studying Cudworth (namely during the early 1680s). We also know that the Cambridge Platonists, and especially Cudworth, were one of the more important influences on Leibniz' thought. Leibniz corresponded with Ralph Cudworth's daughter Damaris who sent him a copy of her father's *True Intellectual System*. This work was accessible to Leibniz especially because extracts from it were published in French in Jean Le Clerc's journal *Bibliothèque Choisie*.[46] Both Leibniz and Locke reject the notion of a plastic nature; but like Cudworth, they both allow for unconscious states of the human soul (in quite different ways, though).[47] As I indicated at the beginning of this paper, the term 'consciousness' became an established one towards the turn of the century and was used extensively not only by Locke, but by a large number of philosophical and theological writers,[48] even though these authors (just like Leibniz and Locke) did not turn to the notion of consciousness as a separate object of inquiry in its own right. In almost all cases the use of 'consciousness' is stripped of the Neo-Platonic context that is present in Cudworth; nevertheless, it is in Cudworth where we find those distinctions and concepts which are essential to the way the notion of consciousness was subsequently applied and discussed.

NOTES

[1] K. J. Grau, for example (*Die Entwicklung des Bewusstseinsbegriffs im 17. und 18. Jahrhundert*, Halle 1916), does not discuss Cudworth at all. B. L. Mijuskovic (*The Achilles of Rationalist Arguments*, The Hague 1974) addresses the issue of 'the unity of consciousness in the 17th and 18th centuries' and discusses the Cambridge Platonists John Smith and, briefly, Ralph Cudworth (pp. 67—70) in that context. However, what Mijuskovic discusses under this title is the argument that a simple, immaterial soul is required to bind our thoughts together. Mijuskovic does not enquire into the *concept* of consciousness itself and how it may be distinguished from other, related concepts. I have made some remarks on Cudworth's notion of consciousness in my German book, *Lockes Theorie der Personalen Identitaet*, Bonn (1983). This paper is a heavily revised and expanded version of those remarks.

[2] Christian Wolff, *Vernuenftige Gedanken von den Kraeften des menschlichen Verstandes und ihrem richtigen Gebrauche in Erkenntnis der Wahrheit*, Halle (1713), I,

par. 1: 'Ich sage aber, dass wir etwas empfinden, wenn wir uns desselben als uns gegenwaertig bewust sind'. See especially Wolff's *Vernuenftige Gedanken von Gott, der Welt und der Seele des Menschen, auch allen Dingen ueberhaupt*, Halle (1719), par. 194; in par. 731—736 the noun 'Bewustseyn' is used: 'Also hebet die voellige Dunckelheit das Bewustseyn auf' (par. 731). For further discussion of issues relating to consciousness see also Wolff's *Psychologia Rationalis*, Frankfurt/Leipzig (1728), and *Psychologia Empirica*, Frankfurt/Leipzig (1732).

[3] *Two Dissertations concerning Sense, and the Imagination. With an Essay on Consciousness*, London (1728). The work has been wrongly ascribed to a certain Zachary Mayne who had died in 1694. There is a recent edition of the *Essay on Consciousness* which contains an important introduction and notes (as well as a German translation of the work) by the editor: Reinhard Brandt (ed.), *Pseudo-Mayne: Ueber das Bewusstsein 1728*, Hamburg (1983).

[4] *A Treatise of the Laws of Nature. By ... Richard Cumberland ... Made English from the Latin by John Maxwell, ... At the End is subjoin'd, An Appendix, containing two Discourses, 1. Concerning the Immateriality of Thinking Substance. 2. Concerning the Obligation, Promulgation, and Observance of the Law of Nature, by the Translator*, London (1727). The full title of the first 'discourse' is *A Summary of the Controversy between Dr. Samuel Clarke's and an anonymous Author, concerning the Immateriality of Thinking Substance*. The appendix has separate pagination; the passages quoted in the text are from p. 5 of the appendix. Cumberland's *De Legibus Naturae* was first published in Cambridge in 1672.

[5] The controversy took place from 1706 to 1708 and was initiated by Henry Dodwell's *An Epistolary Discourse, proving, from the Scriptures and the first Fathers, that the Soul is a Principle naturally Mortal*, London (1706).

[6] This is how Samuel Clarke defined 'consciousness' in *A Second Defense of an Argument made use of in a Letter to Mr. Dodwell*, London (1707), p. 42.

[7] See R. R. Wellman, 'The Question posted at *Charmides* 165a—166c', in *Phronesis* (1964), pp. 107—113. And especially: E. Martens, *Das selbstbezuegliche Wissen in Platons 'Charmides'*, Munich (1973). K. Gloy, 'Platons Theorie der 'episteme heautes' im Charmides als Vorlaufer der modernen Selbstbewusstseinstheorien', in *Kant-Studien* 77, pp. 137—164 (1986).

[8] See for example *De Anima* III, 4 (430a2—5) and III, 2 (425b12).

[9] For questions of etymology and the notions of consciousness and conscientia in ancient thought see Friedrich Zucker, *Syneidesis-Conscientia*, Jena (1928). Gertrud Jung, 'Syneidesis, Conscientia, Bewusstsein', in *Archiv fuer die gesamte Psychologie* **89**, pp. 525—540 (1933). O. Seel, 'Zur Vorgeschichte des Gewissens-Begriffes im altgriechischen Denken', in H. Kusch (ed.), *Festschrift Franz Dornseiff*, Liepzig (1953), pp. 291—319. For the etymology of the English term see especially C. S. Lewis, *Studies in Words*, Cambridge (1960), pp. 181—213; see also the *Oxford English Dictionary* under 'conscience', 'conscious', and 'consciousness'.

[10] See for example Hobbes, *Leviathan* (1651), I, 7; and John Wilkins, *An Essay towards a real Character, and a Philosophical Language*, London (1668), p. 22.

[11] *De Vera Religione*, 39. On Augustine see G. O'Daly, *Augustine's Philosophy of Mind*, London (1987); especially pp. 20—21, 102—105, 148—151, 169—171, 207—211.

¹² *Quaestiones disputatae de Veritate* I, 9.
¹³ R. Goclenius, *Lexicon Philosophicum*, Frankfurt (1613), pp. 247, 379, 971. J. Micraelius, *Lexicon Philosophicum*, Stettin (1662) (second edition; first edition: (1653), column 1208: 'REFLEXUM est, quod in se redit. Sic *actus animi reflexus* est, quo se ipsum intelligit'. See also Grau *op. cit.* p. 105.
¹⁴ *Summa Theologiae* 1 a, 87, 3: '. . . unde alius est actus quo intellectus intelligit lapidem, et alius est actus quo intelligit se intelligere lapidem'; see also 1a, 85, 2.
¹⁵ *Summa Theologiae* 1 a, 87, 3.
¹⁶ Goclenius, *op. cit.*, p. 971: 'Reflexio enim Intellectus eis est, cum postquam intellectus concepit rem aliqua, rursus concipit se concepisse eam, & considerat ac metitur, qua certitudine & modo illam cognoverit,. . .". See also p. 248.
¹⁷ Antony Cade, *A Sermon on the Nature of Conscience* (1621) especially pp. 19—22. See also William Perkins, *A Discourse of Conscience* (1596). William Ames, *Of Conscience with the Power and Cases thereof* (1639) (Latin edition 1630). Henry Hammond, *Of Conscience*, London (1646). Robert Sanderson, *De Obligatione Conscientiae*, London (1647). Jeremy Taylor, *Ductor Dubitantium, or the Rule of Conscience in all her general Measures*, London (1660). For scholastic definitions of *conscientia* see Timothy C. Potts, *Conscience in Medieval Philosophy*, Cambridge (1980). See also Goclenius, *op. cit.*, p. 447; and Micraelius, *op. cit.*, column 321.
¹⁸ Herbert of Cherbury, *De Veritate*, London (1645) (3rd edition; first edition (1624) VI, p. 104. *Conscientia* is said to be that by which we examine what is good and evil, and by means of which we apply the 'common notions' to particular cases.
¹⁹ References to Descartes' works are to the standard edition by Ch. Adam and P. Tannery, *Oeuvres de Descartes*, revised edition, Paris (1964—76) (abbr.: AT) and to *Rene Descartes. Philosophical Writings*. Edited and translated by J. Cottingham, R. Stoothoff, D. Murdoch, 2 Vols., Cambridge (1984—5) (abbr.: *Philosophical Writings*).
²⁰ On 'conscience' and 'conscientia' in Descartes and seventeenth century French thought see Ruth Lindemann. *Der Begriff der conscience im franzoesischen Denken*, Heidelberg (1938). G. Lewis, *Le Probleme de l'Inconscient et le Cartesianisme*, Paris (1950). Wilhelm Halbfass, *Descartes' Frage nach der Existenz der Welt*, Meisenheim am Glan (1968), pp. 90—93, 133—135. Robert McRae, 'Innate Ideas', in R. J. Butler (ed.), *Cartesian Studies*, Oxford, Blackwell (1972), pp. 32—54. Robert McRae, 'Descartes' Definition of Thought', *ibid.*, pp. 55—70. D. Radner, 'Thought and Consciousness in Descartes', in *Journal of the History of Philosopohy* **26**, pp. 439—452 (1988).
²¹ See the replies to the objections to the second *Meditatio* (AT VII, p. 160; *Philosophical Writings*, Vol. 2, p. 113); see also the second *Meditatio* (AT VII, p. 28; *Philosophical Writings*, Vol. 2, p. 19), and the *Principia Philosophiae* I, 9 (AT VIII, p. 7; *Philosophical Writings*, Vol. 1, p. 195). Elsewhere Descartes argues that we know our thoughts through 'conscience or internal testimony' (At X, p. 524; *Philosophical Writings*, Vol. 2, p. 418). In *Philosophical Writings* Descartes' 'conscium esse' is translated as 'to be aware'.
²² AT V, pp. 220—2. See also Anthony Kenny (ed.), *Descartes. Philosophical Letters*, Oxford (1970), pp. 234—5.
²³ 'Conscium esse est quidem cogitare et reflectere supra suam cogitationem', *Descartes et Burman* (AT V, p. 149). See also John Cottingham (ed.), *Descartes' Conversation with Burman*, Oxford (1976), p. 7, and Cottingham's comments on this passage, p. 61.

[24] AT V, p. 149; Cottingham, *op. cit.*, p. 7.
[25] Replies tot the sixth objections to the *Meditations* (AT VII, p. 422; *Philosophical Writings*, Vol. 2, p. 285). For attempts to attribute to Descartes a clear distinction between consciousness and reflection on the basis of this and similar passages see Robert McRae, *Leibniz: Perception, Apperception, and Thought*, Toronto and Buffalo (1976), p. 10, and more recently, Richard E. Aquila, 'The Cartesian and to a certain 'Poetic' Notion of Consciousness', in *Journal of the History of ideas* **49**, pp. 543—562, esp. 546—7 (1988), Aquila does not mention Cudworth.
[26] See Halbfass, *op. cit.*, p. 92.
[27] *Leviathan* I, 2.
[28] AT VII, p. 173; *Philosophical Writings*, Vol. 2, p. 122f. See also Hobbes, *De Corpore* IV, 25, 1 & 5. Spinoza, too, mentions the infinite regress of reflection, but does not seem to regard it as a problem; see *Ethica* II P 21 Schol.
[29] *Traité de l'Esprit de l'Homme*, Paris (1666), p. 54. On de la Forge see Albert G. A. Balz, *Cartesian Studies*, New York (1951), especially p. 95.
[30] Antoine Arnauld, *Des Vrayes et des Fausses Idées*, Paris (1683).
[31] *De la Recherche de la Vérité*, III, 1, 1.
[32] *Ibid.*, III, 2, 7.
[33] Matthew Hale, *The Primitive Origination of Mankind*, London (1677), p. 24.
[34] John Wilkins, *Of the Principles and Duties of Natural Religion*, London (1675), p. 3.
[35] On this 'reflection argument' see Emily Michael and Fred S. Michael, 'Two Early Modern Concepts of Mind: Reflecting vs. Thinking Substance', in *Journal of the History of Philosophy* **27**, pp. 29—49 (1989).
[36] On the notion of plastic nature and its history, especially in the seventeenth century, see W. B. Hunter, Jr.: 'The Seventeenth Century Doctrine of Plastic Nature' in *The Harvard Theological Review* **43**, pp. 197—213 (1950); also G. Aspelin, *Ralph Cudworth's Interpretation of Greek Philosophy*, Goeteborg (1943), pp. 13—15, 26—31; John Passmore, *Ralph Cudworth. An Interpretation*, Cambridge (1951), pp. 19—28; Lydia Gysi, *Platonism and Cartesianism in the Philosophy of Ralph Cudworth*, Bern (1962), pp. 17—24; R. A. Greene, 'Henry More and Robert Boyle on the Spirit of Nature', in *Journal of the History of Ideas* **23**, pp. 451—474 (1962).
[37] See Plotinus, *Enneads* III, 8, 4: Vol. 8, 11/23. For the notion of consciousness in Plotinus see especially H. R. Schwyzer, 'Bewusst' und 'unbewusst' bei Plotin', in *Les Sources de Plotin. Entretiens sur l'Antiquité classique*, Vol. V (Fondation Hardt), Geneva (1960), pp. 343—390; on the term 'synaisthesis' as feeling see pp. 355ff.; see also E. W. Warren, 'Consciousness in Plotinus', in *Phronesis* **9**, pp. 83—97 (1964); and G. J. P. O'Daly, *Plotinus' Philosophy of the Self*, Shannon, Ireland (1973).
[38] On the relationship of Cudworth to Descartes see John Passmore, *op. cit.*, pp. 8—11, 23—28; Lydia Gysi, *op. cit.*; J. C. Gregory, 'Cudworth and Descartes', in *Philosophy* **8**, pp. 454—467 (1933); D. B. Sailor, 'Cudworth and Descartes' in *Journal of the Histsory of Ideas* **23**, pp. 133—140 (1962); on Descartes' impact on seventeenth century English thought in general see now G. A. J. Rogers, 'Descartes and the English' in J. D. North and J. J. Roche (eds.), *The Light of Nature*, Dordrecht (1985), pp. 281—302.
[39] Passmore, *op. cit.*, p. 31, says that for Cudworth 'knowledge is always self-knowledge'.

⁴⁰ EIM 78; see Hobbes, *De Corpore* chapt. 25, 2; *Leviathan* I. 1.
⁴¹ See also EIM 80ff. The Cambridge Platonist John Smith puts forward a similar argument: 'For that which we call sensation, is not the motion or impression which one body makes upon another, but a recognition of that motion; and therefore to attribute that to a body, is to make a body privy to its own acts and passions, to act upon itself, and to have a true and proper self-feeling virtue' ('A Discourse demonstrating the Immortality of the Soul' in *Select Discourses* (1660); quoted from the third edition, London (1821), p. 81).
⁴² EIM 154—5: 'the Intellect . . . raises and excites within it self the Intelligible Ideas of Cause, Effect, Means, End, Priority and Posteriority, Equality and Inequality, Order and Proportion, Symmetry and Asymmetry, Aptitude and Inaptitude, Sign and Thing signified, Whole and Part, in a manner all the Logical and Relative Notions that are'.
⁴³ EIM 119: '. . . the Soul is inwardly Conscious that they [i.e. *Phantasms*] are raised up by its own Activity'.
⁴⁴ Here, the term 'self-conscious' does not refer to a consciousness of *self*, but to the consciousness of one's own *cogitations*.
⁴⁵ For an interpretation of the passage in Plotinus see Schwyzer, *op. cit.*, p. 371f. Plotinus also ascribes this basic self-feeling or 'hoion synaisthesis' to the *One* which cannot, unlike the *Nous*, have self-knowledge, because this would mean that the *One* would be divided into subject and object; see *Ennead* V, 4, 2.
⁴⁶ See Rosalie L. Colie, *Light and Enlightenment. A Study of the Cambridge Platonists and the Dutch Arminians*, Cambridge (1957), pp. 117 ff. Cudworth's 'consciousness' was translated into French as 'conscience' and 'connaissance reflechie' (*ibid.* p. 126).
⁴⁷ The best discussions of Leibniz' notion of consciousness are those by Robert McRae, *Leibniz: Perception, Apperception, and Thought*, Toronto and Buffalo (1976), I add by Mark Kulstad, 'Leibniz on Consciousness and Reflection', in *The Southern Journal of Philosophy*, Vol. 21, Supplement (1983), pp. 39–56.
⁴⁸ See for example John Turner, *A Discourse concerning the Messias . . . to which is prefixed a large Preface, asserting and explaining the Doctrine of the Blessed Trinity, against a late Writer of the Intellectual System*, London (1685). Richard Burthogge, *An Essay upon Reason, and the Nature of Spirits*, London (1694). William Sherlock, *A Vindication of the Holy and Ever Blessed Trinity*, London (1690). William Sherlock, *A Defence of Dr. Sherlock's Notion of a Trinity in Unity*, London (1694). Robert South, *Animadversions upon Dr. Sherlock's Book, Entituled a Vindication of the Holy and ever Blessed Trinity*, London (1693). Robert South, *Tritheism Charged upon Dr. Sherlock's new Notion of the Trinity*, London (1695). John Sergeant, *Solid Philosophy Asserted*, London (1697). For a discussion of the notions of consciousness in these works see my *Lockes Theorie der personalen Identität*, Bonn (1983), pp. 106–107 (Turner), pp. 76–78 (Burthogge), pp. 107–116 (Sherlock and South), pp. 121–3 (Sergeant). My main discussion of Locke's conception of consciousness is on pp. 88–104.

ALEXANDER JACOB

THE NEOPLATONIC CONCEPTION OF NATURE IN MORE, CUDWORTH, AND BERKELEY

The Neoplatonic conception of Nature was one of the major bulwarks against the threat of atheism posed by the materialistic and mechanistic philosophies of the early modern age. The efforts of Henry More, the Cambridge Platonist, to combat the systems of philosophers like Hobbes and Descartes with his doctrine of the Spirit of Nature may be considered the most original and elaborate adaptation of Neoplatonic metaphysics to the physics and physiology of the seventeenth century.[1] More's theories concerning the Spirit of Nature were repeated shortly after by his fellow Cambridge Platonist, Ralph Cudworth, in his discussion of the 'plastic nature' in *The True Intellectual System of the Universe* (1678), a book which in fact seems to have been the source of Berkeley's ruminations on the same subject in his late antiquarian work, *Siris* (1744). Although these three philosophers agree in their vitalistic interpretation of nature, I think that careful investigation will reveal subtle differences in their understanding of the exact relationship between the intelligible and sensible realms, and the particular manner in which the Divine Mind rules the physical universe as Providence.

As the major source of the natural philosophical notions of the Cambridge Platonists is Plotinus, we will do well to survey Plotinus' conception of nature in the *Enneades*.[2] The central concept in the Plotinian system is that of Absolute, the One from which all arises and to which all must return. The One is beyond being and quite beyond all duality of act or definition. But its infinite goodness is irradiated as Mind, which in turn produces the phenomenon of matter through its energy, or Soul. Thus the intelligible realm of Plotinus is constituted of three hypostases, the Primal One, the Mind, and the Soul. Matter is in itself non-being and merely the limit of irradiation of the Mind, or, to express it differently, it is the term of spiritual reflection and consequently, a metaphysical mirror-image of the Mind. The hypostases are all aspects of the One and the result of contemplation, θεωρία. θεωρία is a term well chosen for it suggests that the hypostases are evolved by an essential contemplation that is simultaneously a positing of that essence at a lower level of being. The One, in regarding itself, thus posits the Mind, and the Mind, in contemplating its source posits itself as the energetic Soul. Similarly, the Soul posits itself in external manifestation as the material universe, which according to Plotinus, is 'a shadow as broad as the Reason Principle proceeding from the Soul' (IV, iii, 9).[3] Since the last emanation is into the nothingness of matter, Nature as an entity has only a slight individuality and shadowy existence. It is, in fact, the lower phase of the Soul at the moment that it looks outwards to its limit of contemplation, matter. As Plotinus puts it, '[Nature] gazes and the figures of the material world take being as if they fell from its contemplation' (III, viii, 4).

Nature in general is the vegetative soul and in humans, therefore, the lowest part of their personality, below the sensitive and rational. The operations of Nature are not conscious but unconscious:

> For the Vision which Nature broods inactive, is a self-intuition, a spectacle laid before it by virtue of its unaccompanied self-concentration ... It is a Vision silent but somewhat blurred, for there exists another [i.e. Soul] of which Nature is the image: hence all that Nature produces is weak (III, viii, 4).

Again, in IV, iv, 13, Plotinus points to the difference between the illuminated activity of the Soul and the dimmer reflection that is Nature:

> But what is the difference between the Wisdom thus conducting the universe and the principle known as Nature? This Wisdom is a first (within the All-Soul) while Nature as an image of that Wisdom, and as a last in the Soul, possesses only the last reflection of the Reason Principle ... Nature, thus, does not know, it merely produces: what it holds, it passes, automatically to its next, and this transmission to the corporeal and material constitutes its making power.

It is important to note that although the perception of Nature is as unconscious as that of sleep (τοῦ ὕπνου, III, viii, 4), its mode of production is effortless and organic since it does not have to search for an object that it does not already possess: 'Nature does not lack: it creates because it possesses' (III, viii, 3). Unlike mechanical processes, natural ones are conducted by virtue of the Ideas instinct in the higher phase of the Soul:

> In sum, then, the Intellectual Principle gives from itself to the Soul of the All which follows immediately upon it; this again gives forth itself to its next, illuminated and imprinted by it; and that secondary Soul at once begins to create, as under order,

unhindered in some of its creations, striving in others against the repugnance of matter (II, iii, 17).

Since the entire phenomenal universe is thus informed by Soul, albeit at a secondary level, it partakes of the unity of the intelligible realm. This is evidenced in the sympathetic unity of its parts which accounts for the efficacy of prayers, magic spells, and astral influence. And it is also the means by which the Divine Law is manifested as Nemesis in the natural world:

> ... linked as we are by affinities within us towards the answering affinities outside us, becoming by our soul and the conditions of our kind thus linked — or better, being linked by Nature — with our next highest in the celestial or daemonic realm, and thence onwards with those above the Celestials, we cannot fail to manifest our quality ... Anyone that adds his evil to the total of things is known for what he is, and, in accordance with his kind, is pressed down into the evil which he has made his own, and hence, upon death, goes to whatever region fits his quality — and all this happens under the pull of natural forces (IV, iv, 45).

It is important to note the stress on natural forces as the agent of Providence. For, Plotinus cannot attribute any evil, however indirectly, to the intelligible sphere. As he reiterates,

> necessities inherent in the nature of things account for all that comes from above — Secondly, there is the large contribution made by the individual. Thirdly, each several communication, good in itself, takes another quality in the resultant combination. Fourthly, the life in the Cosmos does not look to the individual but to the whole. Finally, there is Matter, the underlie, which being given one thing receives it as something else, and is unable to make the best of what it takes (IV, iv, 39).

Thus, he concludes, 'If all this is sound, at once our doubts fall and we need no longer ask whether the transmission of any evil is due to the gods'.

The Plotinian model of the intelligible and sensible universe was revived in the seventeenth century with both accuracy and originality in the philosophy of Henry More. More's views on spirit and matter are most clearly presented in his first major philosophical work, *A Platonick Song of the Soul* (1647),[4] since it displays the various entities of the spiritual and material spheres in their original Neoplatonic form, without the Cartesian terminology of 'substances' which has led many to consider More as a typical 'dualist'.[5] The two realms are represented by More in a 'universal Ogdoas' constituted of eight concentric orbs called Ahad (God, or the One), Aeon (Intellect), Psyche (Soul), Semele (Intellectual Imagination), Arachne (Sense-Perception), Physis (Nature), Tasis (Physical Extension), and Hyle (Matter), arranged in the following manner:[6]

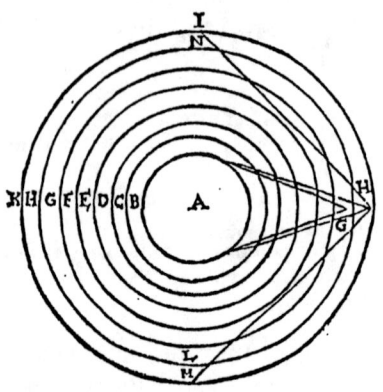

The first four orbs belong to the intelligible realm, while the latter four are sensible phenomena. But the four levels in each realm exactly mirror the corresponding four levels in the other. So that, God is the opposite of Hyle, Intellect the opposite of Tasis, Psyche that of Physis, and Semele that of Arachne. The crucial elements in this scheme are Psyche and Physis, or the Soul and the Spirit of Nature (Semele and Arachne are merely instruments used by the Soul and the Spirit of Nature respectively). The world of phenomena is formed by Psyche's movement outwards, which causes the ideas contained in her consort, Aeon's mind to take on the appearance of physical extension, or Tasis. If not for this Psychic movement out of its Divine centre as if there were something outside it, there would be nothing but the Deity. But the energy of the Divine Soul will forever manifest itself in the imaginary medium of matter. The movement of the Soul is represented in the figure by the cone formed by the infinite spiritual base A and the cusp of the mathematical point G to which spirit contracts in its 'audacious ramble'[7] outwards to matter. This movement of spirit between the real infinity of God (A) and the reflected one of matter (H) is instantaneous. As More expresses it poetically:

> Lo! here's the figure of that mighty Cone
> From the straight Cuspis to the wide-spread Base
> What's infinitely nothing here hath place
> What's infinitely all things steddy stayes
> At the wide Basis of this Cone inverse
> Yet its own essence doth it swiftly chace
> That motion here's no motion. (Antipsychopannychia, I, 9)

In the process of manifestation, the Spirit of Nature works as the agent of the Divine excellencies on matter. As the Spirit of Nature is spirit tending towards matter and as the realm of matter is the obverse of spirit, the Spirit of Nature is the most primitive stage in the manifestative of spirit and does not possess reason as fully individual souls do. This is the reason why it is unconscious and is 'sometimes puzzeld and bungells in ill disposed matter'.[8] More further calls it 'not the divine Understanding it self, but ... an Artificer's imagination separate from the Artificer, and left alone to work by it self without animadversion.'[9]

When we come to More's later elaboration of the conception of the Spirit of Nature in his prose treatise, *The Immortality of the Soul* (1659, rev. 1662, rev. in Latin 1675), we find that it is part of a more carefully defined doctrine of substances. In Book I, chap. 3, he defines matter as 'a substance impenetrable and indiscerpible' and spirit as 'a substance penetrable and indiscerpible', endued with 'self-motion, self-penetration, self-contraction and dilatation, and the power of moving matter' (chap. 7).[10] This apparently dualistic definition of spirit and matter should not hinder us from noting that the distinguishing characteristic of matter is still 'impenetrability' or 'resistance' to spirit. Besides, both spirit and matter are extended, though spiritual extension is 'essential' in contradistinction to the 'integral' extension of matter (Preface, Sec. 3, and Book I, Chap. 6, Sec. 6). This quality of 'spissitude' or 'self-reduplication' allows spiritual extension to be at once infinite and eternal, whereas material extension is defectively compounded of divisible parts.

The created spirits in the universe are divided into four main species, the souls of angels, human souls, animal souls, and seminal forms (Chap. 8). All these types of spirit participate in the Spirit of Nature, but it is the last item, or λόγοι σπερματικόι, that offers the clearest example of the Spirit of Nature in its most elementary form: 'A Seminal Form is a created Spirit organizing duly-prepared Matter into life and vegetation proper to this or the other kind of Plant'. When More comes to the Spirit of Nature itself in Book III, Chap. 12, we are not surprised to find that its definition is but an enlargement of that of a seminal form:

> The Spirit of Nature — is A Substance incorporeal, but without Sense and Animadversion, pervading the whole matter of the universe, and exercising a Plastical power therein according to sundry predispositions and occasions in the parts it works upon, raising such Phaenomena in the World as cannot be resolved into mere mechanical powers.

The crucial feature of this definition is the focus on the power of the Spirit of Nature to move and direct the matter within which it operates. In his Notes to the third edition of *The Immortality of the Soul*, More significantly states that

> The vital Motion which all the Matter of the World hath, is from the Spirit of Nature, and also governed by it, unless so much as is permitted to particular souls to excite in it, which may direct this Motion of the Matter to their own Purposes, but not to such a degree as they please (Notes on Book II, Chap. 8, Sec. 10).

When we recall More's original conception of the dynamic relationship between God and matter in the *Platonick Song of the Soul* (See above p. 104) we realize that the vital motion of the Spirit of Nature is essentially a dynamic reflection of the order of God's Mind in the formless nothingness of matter. This is the reason why, in all of the operations of the Spirit of Nature as described by More, we notice its unconsciousness, its merely vegetative and sensitive capacities, and its inequality to reason. The Mind is always considered to be entirely distinct from the sensible world. This is made clear in More's account of the reason why the Spirit of Nature expresses itself in diverse forms. For, he says,

> it contains vitally in every point of it all Mundane Forms whatever, with all the particular parts of everyone of them, (for the omniform Life of the Spirit of Nature, for so much as concerns the World, answers to the omniform divine Intellect it self) (Notes on Book II, Chap. 12, Sec. 12).

The difference between the two realms and the imaging correspondence between them are focussed in the phrase 'answers to'. It is interesting to note that Richard Burthogge, in his *Essay upon Reason and the Nature of Spirits* (1694), specifically refers to the distinction that More makes between the Spirit of God and the Spirit of Nature and, in disagreeing with More's view, reveals the Scriptural basis of his own immanentist philosophy. He declares that More maintains

> that God doth actuate all the Matter of Natural Corporeal World by the Spirit of Nature; but that he actually acts in and governs the world of Men and Angels by the Spirit of God. But I have shewed already from the Scriptural *Hypothesis*, that it is *one* Spirit, [the *Mosaical*] that Actuates, and Acts in All, in Men and other Animals, as well as in the World of meer Nature, as to all the operations commonly called Natural . . . In fine, the *Principium Hylarchium* or Spirit of Nature (as this Learned Person calls it,) is but a *Plastick Faculty*, of the *Mosaical* Spirit.[11]

Burthogge's tendency to closely identify the World-Soul with the Spirit of Nature is evident also in the system of Cudworth, and heightened in that of Berkeley, as I shall show below. More recent commentators on More have, unfortunately, failed to understand this crucial distinction in his metaphysical system and have, consequently, tended to misinterpret his philosophy.[12]

To proceed with More's system, when the Spirit of Nature is individuated into the several souls, human and animal, they possess their own plastic faculty, which, imitating the formative action of its cosmic source, works at first in the embryo and then gradually develops the physical structure of the body until its mature state is reached. More even surmises that, at the very earliest stages in the formation of the embryo, the 'first rude preparative strokes' may be given by the Spirit of Nature itself:

> But the *punctum saliens*, or Life-point, discovers not any proper sense, but only Life when it withdraws it self from any hurtful touch: as rather do what we call the Plant-Animals; which I do not take to be any proper Soul, but through the Spirit of Nature, which hath either no Sense at all, or what is very dull. (Notes on Book II, Chap. 10, Sec. 12)

The particular process by which the plastic part of the soul forms the embryo is described by More thus:

> the Plastick of the Soul of Man contains also vitally in every point, as it were, of it, the whole humane Form enlarged to its just Bigness, with all the Degrees, Parts, or Particles of it: So that none need wonder, that either the Plastick of the Spirit of Nature or of the humane Soul, should discharge its Plastical Office through the whole Formation of the Foetus, with all the degrees of the Increase in it: For that Homini-Form Life being every where in the Plastick of the Soul is exerted as occasion is according to all its Parts and Degrees; and the Parts contracted within themselves, conspire all the same way, and impress themselves throughout the whole Process the same Form on Matter; but

greater and greater still in all the Parts, according to the Body's Growth or Increase, until it comes unto its just Stature (Notes on Book II, Chap. 12, Sec. 12).

More is thus a preformationist rather than an epigeneticist. Impelled by the manifold forms and the Laws implanted in it, the Spirit of Nature carries out its work of fabrication 'by a very sure, tho' blind instinct, and by such an one as is vital, not intellectual — and that also in the general, which is always for the best' since it acts 'as from the Pattern of Divine Counsel and Wisdom' (Notes on Book II, Chap. 11, Sec. 8).

Significantly, More illustrates this manner of operation by the analogy of the working of the vital plastic faculty of human souls during sleep:

which we ought the less to wonder at, since we ourselves excite Phantasms without any previous perception; but we perceive them to be excited through a certain Life, that is intimately in our Souls and previous unto any operations: as appears especially in Dreams, where the vital Substance of the Soul it self affected by either Choler, or else by a sanguine or melancholic humour, doth not by any Counsel, but vitally and necessarily, without any previous perception, excite suitable scenes of things, and perceives them when so excited in it. Why may not therefore the vital Spirit of the World as congruously be conceived to raise Plasmata, or Works, without any previous Perception, or to move and dispose Matter according unto certain Laws, as our Souls without previous Perception to raise Phantasms. (*Ibid.*)

He further adduces the phenomenon of habits, intellectual as well as plastic. For,

the Perceptive part of the Soul it self — even in the use of intellectual habits, produceth readily its Notions or Apprehensions of things, rather by a vital intellectual operation, than by Perception: For neither doth it perceive them before it hath produced them; nor doth it then produce them by a previous Perception, but by a sort of immediate vital Intellectual faculty of the Soul; for otherwise the Process would be infinite. In the vital Essence therefore, whether intellectual or plastical, is the Habit rooted. (*Ibid.*)

The most concerted defence of the operations of the Spirit of Nature occurs in More's long discussion of instinctual life in Book III, Chap. 13. He begins by reminding us of the axiom 'Frustra fit per plura quod fieri potest per pauciora.' So it is logical that one single vegetative spirit be responsible for the common life of all plants whatsoever. This explains 'their Transmutations into other Species, the growing of Slips, and the like. For there is one Soul ready every where to pursue the advantages of prepared Matter. Which is the common and only λόγος

THE NEOPLATONIC CONCEPTION OF NATURE 109

σπερματίτης of all Plantal appearances' (Book III, Chap. 13, Sec. 7). Only beasts and men have been endowed with individual souls that allow them to enjoy the sensuous world.

In the animal world, the Spirit of Nature manifests itself as instinct, for, indeed, any regular natural phenomenon is due to it. The different forms of instinctual behaviour include the building of nests by birds in such a way as to protect the eggs from intruders. This is especially evident in the case of indocile birds which build the most intricate nests. The cocoon of the silkworm is even more significant in that it is not only for 'the accomodating of the Individual, but a plot for the propagation of the Species' — which is clearly the concern of a larger agent than any bestial one can be:

> For if there be an Impulse from an Extrinsecal Principle upon any particular Animal, it is most sure to be then, when that Animal is transported from the pursuance of its own particular accomodation to serve a more publick end. For from whence can this motion be so well as from that which is not a particular Being, but such as in whose Essence the scope and purpose of the general good of the world and of all the Species therein is vitally comprised, and therefore binds all Particulars together by that common Essential Law which is it self, occasionally impelling them to such actions and services (either above their knowledge or against their particular Interests) as is most conducive to the Conservation of the Whole? (Book III, Chap. 13, Sec. 9)

The pervasiveness of the Spirit of Nature accounts for the sympathetic unity of the universe, a notion familiar from Plotinus' psychology:

> For the Spirit of Nature in a sort activates and informs all Bodies whatsoever: Whence is easily understood this Community and Consent of Life I speak of, and also real Sympathy. (Notes on Book III, Chap. 12, Sec. 4)

The phenomena which More adduces to demonstrate the vital unity effected by the Spirit of Nature include the sympathetic vibration of musical strings, sympathetic cures and woundings, the formation of signatures in the womb, gravity, and magnetism (Book III, Chap. 12).

The way the Spirit of Nature effects the dictates of Divine Providence in the realm of animal and human souls is by 'transporting ... particular Souls and Spirits in their states of silence and Inactivity to such Matter as they are in a fitness to catch life in again' (Book III, Chap. 13, Sec. 10). The transportation of souls is conducted in the same manner as magnetic particles are conducted from one pole to the other, that is, by virtue of the sympathetic unity of the universe produced by the pervasive substance of the Spirit of Nature. If the soul has been virtuous in its terrestial condition, it will be borne to an aerial one; if not, it will return to a human. Sometimes souls descend from the higher aerial state to an earthly one, which is a special proof of More's favorite theory of the preexistence of the soul. The movement from an aerial vehicle to an earthly one [13] is accompanied by forgetfulness of its past actions, since 'amidst the moistures of the Womb it lives the Life of a Plant, rather than an Animal' (Notes on Book III, Chap. 13, Sec. 3). This is because the first expression of the Spirit of Nature or its individual counterpart, the plastic faculty of the soul, is as a dull vegetative life. However, since the human souls that are borne by the Spirit of Nature possess also sensitive and, later, receive rational faculties, the latter too will eventually manifest themselves.

It is important to note that although More does not very sharply differentiate the sensitive from the plastic soul, he does clearly distinguish the rational from the plastic. This division of the soul into an intellectual and a plastic/sensitive part is accentuated in Book III, Chap. 1, Sec. 2, where he declares that the soul's intellectual operation will persist in the aerial vehicle, its conjunction to the bodily matter in its terrestial form 'in all likelihood engaging onely the Plastic and Sensitive Powers of the Soul'. In other words, the rational soul is believed by More to be of such a superior status that even though its ethereal element is contained in the animal spirits which are the instrument of the plastic part of the individual soul, yet it is never totally merged with the animal spirits. In fact, he follows Aristotle in maintaining that the embryo does not have a rational soul at the earliest stages of its formation, but receives it a little later. Furthermore, the rational soul is not only distinct from the lower but also opposed to it. In the *Enchiridion Ethicum* (1666), for example, he declares that 'the Intellectual part of the Soul strives with the Plastic' (Book I, Chap. 6, Sec. 8).

The distinction of the rational from the lower soul is of crucial importance because it allows More to free the Divine Mind from the responsibility for any of the imperfections of Nature. By conceiving of the Spirit of Nature as the lower aspect of the Soul operating on matter, More was able to accurately describe not only the vital organic processes of the universe but also its unconscious mode of operation. This unconsciousness of Nature is explained by the peculiar imaging of the intelligible world in the sensible whereby the last projection of the Soul appears as the first stirrings of Reason in matter.

When we turn to the doctrine of 'plastic nature' discussed in Cudworth's *True Intellectual System of the Universe*, we find many of More's notions of the Spirit of Nature repeated in a compendium of similar notions of Nature culled from both ancient philosophers and modern. But Cudworth's theory of the plastic nature does not occupy an integral position in a metaphysical model of the universe, as it does in More's. It is significant that the major discussion of the plastic nature occurs as a 'Digression' at the end of the third chapter of the *Intellectual System*, whereas his ontological discussions of spirit and matter follow much later in the fifth. According to Cudworth's ontological system, spirit and body are distinguished by impenetrability and activity. In this, Cudworth follows More's definitions in the *Immortality of the Soul*. However, unlike More's spirit, Cudworth's spirit is not definitely stated to be an extended incorporeal substance, but merely a centre of energy or self-activity. In fact, he admits to being uncertain whether spiritual substance is extended or not: 'But whether this substance be altogether unextended, or extended otherwise than body; we shall have every man to make his own judgement concerning it.'[14] Although Cudworth considers this difference insignificant, it is indeed not so. For it is hard to understand what the plastic nature is unless it be first established that spirit is extended in such a way that in its farthest reaches, that is, next to the reflecting points of matter, it appears bereft of all consciousness that characterized it within its original Intellectual sphere.

Cudworth also considers his two substances to be both created and finite. Yet they are possessed of 'virtual infinity', which in the case of corporeal substance means the possibility of indefinite increase of magnitude, for

How vast so ever the finite world should be, yet is there a possibility of more and more magnitude and body, still to be added to it, further and further, by divine power infinitely, or the world could never be made so great, no not by God himself, as that his own omnipotence could not make it yet greater.[15]

In the case of incorporeal substance, it means the 'potential omniformity' of the soul to imagine all things as well as the occasional transcendance of the soul above the limits of reason. In other words, Cudworth's conception of spirit is mostly restricted to the innate energy in thinking or moving beings, and does not relate to the infinite energy of God Himself diffused as Divine Space, as More described it in his *Enchiridion Metaphysicum*, Cap. 8, Secs. 8ff. Interestingly, Cudworth does at one juncture, consider space as 'a nature distinct from body, and positively infinite', and that it must be the 'affection' of some infinite incorporeal substance, namely God.[16] But he does not relate this infinite incorporeal substance to the soul, which is characterised more particularly as life and cogitation. Besides, the 'vital motion' of Cudworth's spiritual substance is never explained since its special mode of activity is not either.[17] More, on the other hand, having first described the energetic extension or spissitude of the soul, could identify spirit with reified motion, as he did in *The Immortality of the Soul*, and the *Divine Dialogues* (1668).[18] Since Cudworth has no clear notion of the extension of spirit and can only conceive of it in quasi-Aristotelian fashion as the inner motivating force of a particular body, it is not surprising that he has no conception of a universally extended world-soul either. But the Spirit of Nature is the obverse of the *anima mundi*, or the latter seen from the point of view of matter, so that the one cannot exist without the other. When Cudworth introduces his plastic nature, it is indeed as an ill-explained *tertium quid*, or 'third man' as Passmore terms it,[19] that 'is inserted into things themselves' in order to serve as the 'servant and executioner of Providence'.[20]

Cudworth's main concern in positing the plastic nature is to show the insufficiency of mechanical agency to explain organic productions. But he can only conceive of it as a vital substance which 'together with mechanism ... runs through the whole corporeal universe' (Book I, Chap. 3, Sec. 37, Art. 3). Given this restricted understanding of the status of the plastic nature, it is not clear how Cudworth can reconcile his view of it as unconsious with his view of it as a sign of the working of a Divine Mind. He represents it as if it were a personal agent of a personal God:

it may well be concluded that there is a Plastick Nature under him, which, as an Inferior and Subordinate Instrument, doth Drudgingly Execute that Part of his Providence which consists in the Regular and Orderly Motion of Matter (*Ibid.*, Art. 5).

We may compare this with More's subtler understanding of the Spirit of Nature as the *imagination* of the Divine Artificer (See above p. 105). Besides, God is not always absent from the processes of natural production, according to Cudworth:

there is also, besides this, a Higher Providence to be acknowledged, which presiding over it, doth often supply the Defects of it, and sometimes Over-rule it; forasmuch as this Plastick Nature cannot act Electively, nor with Discretion (*Ibid.*).

THE NEOPLATONIC CONCEPTION OF NATURE 113

In order to obtain support for his theory, Cudworth surveys the philosophy of the ancients and discerns the notion of the plastic nature in most of them. For instance, he interprets Empedocles' theory of φιλία and νεῖκος as but an obscure exposition of the plastic nature of the world which 'opposing the parts to one another, and making them severally indigent, produces by that means war and contention. And therefore, though it be one, yet, notwithstanding, it consists of different and contrary things ... the seminary and plastic nature of the world may be fitly be resembled to the harmony of disagreeing things'. Heraclitus' 'ethereal body', too, was the same thing as the plastic nature, while Zeno crystallized it in his definition of nature as 'a habit moved from itself, according to spermatic reasons or seminal principles, perfecting and containing those several things, which in determinate times are produced from it, and acting agreeably to that from which it was secreted'. Even the modern Paracelsian chemists bear witness to such a principle in their notion of the 'Archeus' indwelling in all animal bodies (*Ibid.*, Art. 6). From this collection of philosophical antecedents, Cudworth arrives at the conclusion that the plastic nature is to be recognized as the cause of the 'orderly, regular and artificial Frame of Things in the Universe'. And since 'Mind and Understanding is the only true Cause of Orderly Regularity ... he that asserts a Plastick Nature asserts Mental Causality in the World' (*Ibid.*, Art. 7).

We observe that whereas More had focussed on the fact that the Spirit of Nature was without sense or animadversion, Cudworth is more concerned to establish the Mind behind the orderly operations of the plastic nature. Cudworth does describe Nature as '*ratio mersa et confusa*, reason immersed and plunged into matter, and as it were, fuddled in it, and confounded with it'. But we are not told how this plunging of reason into matter exactly comes about. He quotes the *Enneades*, IV, iv, 13, to show that nature is unconscious (see above p. 102) and gives the examples of habits, instincts and dreams that More had already adduced to prove this. He even considers the possibility of a universal plastic nature:

Besides this plastick nature, which is in animals, forming their several bodies artificially, as so many microcosms or little worlds, there must be also a general plastick nature in the macrocosm, the whole corporeal universe, that which makes all things thus to conspire every where, and agree together into one harmony (*Ibid.*, Art. 23).

But, typically, this remains a vague suggestion unsupported by any precise notion of spiritual extension or of a world-soul. Thus, in discussing Aristotle's notion of a mundane soul or Moved Mover, which is beneath the Unmoved Mover, and cause of the vital movements of the universe, Cudworth rather naively remarks 'though there were no such Mundane soul, as both Plato and Aristotle supposed, distinct from the Supreme Deity, yet there might notwithstanding be a Plastick Nature of the universe depending immediately upon the Deity itself' (Art. 24). But the World-Soul is a crucial category in Neoplatonic metaphysics for it alone can explain the dynamic movement between Divine Mind and matter. And that Cudworth does not really understand this is made clear by his reservations in Ch. IV against Plotinus' notion of the consubstantiality of the human soul with the world-soul in Chap. IV,[21] where he thinks that the world-soul that Plotinus talks of must be the same as that which forms the physical world, thereby 'bringing down the third hypostasis of [the Platonists'] trinity so low, and immersing it so deeply into the corporeal world, as if it were the informing soul thereof, and making it to be but the elder sister of our created souls'. He corrects himelf a little later by recognizing that Plotinus did not make the third hypostasis the immediate informing soul of the corporeal world but 'a higher separate soul, or superiour *Venus*, which also was the *Demiurgus*, the *maker*, both of our souls and of the whole world'. But he does not understand the relationship between the two souls in microcosm and macrocosm in Plotinus' system, and consequently thinks that Plotinus must have been 'drousily nodding all the while' when he asserted that the human soul was a part of the world-soul.

Finally, when Cudworth discusses the basic elements of being, we note that he not only distinguishes 'Resisting or Antitypous Extension' and 'Life' (i.e. Internal Energy and Self-activity), but also further subdivides life into conscious and unconscious, the latter being 'plastic nature' (*Ibid.*, Art. 16). Here is revealed the basic weakness of the Cudworthian metaphysical system, for he does not relate the latter form of life to antitypous element in any integral way. It is not surprising, therefore, that Cudworth's final definition of the plastic nature too is marked by an unsatisfactory vagueness:

[The plastick nature] is either a Lower Faculty of some Conscious Soul, or else an Inferior kind of Life or Soul by itself; but essentially depending upon an Higher Intellect (Art. 26).

The vitalistic conception of natural operations that we have studied in

More and Cudworth is repeated in Berkeley's *Siris*. The historical method that Berkeley adopts of pointing to cognate notions of Nature amongst ancient philosophers as well as his references to 'Doctor Cudworth' in this work (Secs. 251, 352) reveal his close reading of *The True Intellectual System of the Universe*. However, although *Siris* is a Neoplatonic work insofar as it attempts to rise from the world of sensible phenomena to the intelligible, it is in fact more Neostoic than Neoplatonic. Cudworth's indifference towards the Soul of the World is in Berkeley transformed into an identification of the Soul with the corporeal aether which curiously performs the same offices as the incorporeal Spirit of Nature or plastic nature of the Cambridge Platonists. Although Berkeley agrees with the Neoplatonic opposition of spirit to matter, the one being characterised by thought and action and the other by resistance (Sec. 290), he does not devote much attention to spirit as a substance. And though there are vague references to the 'vital spirit of the world', Berkeley does not subscribe to the concept of an incorporeal and universal Spirit of Nature. He is more concerned with the first instrument of the Divine Mind, the subtle, invisible, but corporeal aether:

This aether or pure invisible fire, the most subtle and elastic of all bodies, seems to pervade and expand itself throughout the whole universe ... This mighty agent is everywhere at hand, ready to break forth into action, if not restrained and governed with the greatest wisdom. Being always restless and in motion, it actuates and enlivens the whole visible mass, is equally fitted to produce and to destroy, distinguishes the various stages of nature, and keeps up the perpetual round of generations and corruptions, pregnant with forms which it constantly sends forth and resorbs. So quick in its motions, so subtle and penetrating in its nature, so extensive in its effects, it seemeth no other than the vegetative soul or vital spirit of the world (Sec. 152).[22]

The last sentence reveals Berkeley's rather Stoic view of the world, for the spirit of the world or *anima mundi* is considered to be the same as the corporeal aether. In commenting on the motion manifest throughout nature, too, Berkeley combines aether with spirit as if they were one entity:

All those motions, whether in animal bodies or in other parts of the system of nature, which are not effects of particular wills, seem to spring from the general cause with the vegetation of plants — an aethereal spirit activated by Mind (Sec. 258).

However, Berkeley deprives the aether of all 'force' and treats it as being merely 'capable of motion, rarefaction, gravity, and other qualities of bodies' on account of the operation in it of the Intellectual Agent or Mind:

Force or power, strictly speaking, is in the Agent alone who imparts an equivocal force to the invisible elementary fire, or animal spirit of the world, and this is the ignited body or visible flame, which produceth the sense of light and heat. In this chain the first and last links are allowed to be incorporeal: the two intermediary are corporeal, being capable of motion, rarefaction, gravity, and other qualities of bodies (Sec. 220).

The diversification of the spirit of the world into its manifold constituent parts is also explained by Berkeley as being due to the different forces acting in the aether:

The pure aether or invisible fire contains parts of different kinds, that are impressed with different forces, or subjected to different laws of motions, attraction, repulsion, and expansion, and endued with divers distinct habitudes towards other bodies. These seem to constitute the many various qualities, virtues, flavours, odours, and colours which distinguish natural productions. The different modes of cohesion, attraction, repulsion, and motion appear to be the source from whence the specific properties are derived, rather than different shapes or figures ... The original particles productive of odours, flavours, and other properties as well as of colour, are, one may suspect, all contained and blended together in that universal and original seminary of pure elementary fire; from which they are diversely separated and attracted by the various subjects of the animal, vegetable, and mineral kingdoms, which thereby become classed into kinds, and endued with those distinct properties which continue till their several forms, or specific proportions of fire, return into the common mass (Sec. 162).

This might seem akin to the Plotinian notion of Ideas in the higher aspect of the Soul. But not only does Berkeley not subscribe to an *anima mundi* distinct from the aether, but he also declares that Intellect is something *superadded* to the aethereal spirit:

Intellect, superadded to aethereal spirit, fire, or light, moves, and moves regularly, proceeding in a method, as the Stoics, or increasing and diminishing, as Heraclitus expressed it (Sec. 229).

It is significant that Berkeley disagrees with Newton's speculations on the aether in the *Opticks* (Book III, qu. 18, 31), for he finds Newton's conception of the aether as subtler substance than light an 'uncouth explanation':

To explain the vibration of light by those of a more subtle medium seems uncouth explanation. And gravity seems not an effect of the density and elasticity of aether, but rather to be produced by some other cause — Should not therefore gravity seem the original property and first supposed? On the other hand, if force be considered as prescinded from gravity and matter, and as existing only in points or centres, what can this amount to but an abstract, spiritual incorporeal force? (Sec. 225)

Newton had in his varied ruminations on the aether, finally approached something like More's conception of spirit or incorporeal substance for he refers to its 'dilating' and 'contracting' and 'vibrating Motion'.[23] But Berkeley is impatient with any discussion of the quality of spiritual substance, individual or universal. He considers forces as abstractions:

But what is said of forces residing in bodies whether attracting or repelling is to be regarded only as a mathematical hypothesis, and not as anything really existing in nature (Sec. 234).

whereas in More force exists really as a property of the Spirit of Nature. The reason for Berkeley's avoidance of speculations regarding the Soul of the World or its lower aspect, More's Spirit of the Nature, is that he will not have the Deity too far removed from the course of the world. In his view, the Divine Mind that presides over the movements of nature is never a distant First Cause but an agent of continuous providential action:

Nor will it suffice from present phenomena and effects through a chain of natural causes and subordinate blind agents, to trace a divine Intellect as the remote original cause, that first created the world, and then set it agoing. We cannot make even one single step in accounting for the phenomena without admitting the immediate presence and immediate action of an incorporeal Agent, who connects, moves, and disposes all things according to such rules and for such purposes as seem good to Him (Sec. 237).

We note immediately that Berkeley does not subscribe to the unconscious Spirit of Nature as a metaphysical entity. He even contradicts Cudworth's description of Nature as 'ratio immersa et confusa', for according to him:

the formation of plants and animals, the motion of natural bodies, their various properties, appearances, and vicissitudes, in a word, the whole series of things in this visible world, which we call the course of Nature, is so wisely managed and carried on that the most improved human understanding cannot thoroughly comprehend even the least particle therof (Sec. 255).

Significantly, he also explains habits as not unconscious phenomena, but as proceeding from something that understands the rule [of playing on a musical instrument, for example]. Therefore if not from the musician himself, from some other active intelligence, the same perhaps which governs bees and spiders and moves the limbs of those who walk in their sleep (Sec. 257).

He is also against the theory of preformation supported by More, as it precludes the constant intervention of God in organic growth:

Others suppose God did more at the beginning, having then made the seeds of all vegetables and animals, containing their solid organical parts in miniature, the gradual filling and evolution of which, by the influx of proper juices, doth constitute the generation and growth of a living body. So that the artificial structure of plants and animals daily generated requires no present exercise of art to produce it, having been already framed at the origin of the world, which with all its parts hath ever subsisted, going like a clock or machine by itself, according to the laws of nature, without the immediate hand of the artist. But how can this hypothesis explain the blended features of different species in mules and other mongrels? Or the parts added or changed, and sometimes whole limbs lost, by marking in the womb? Or how can it account for the resurrection of a tree from its stump, or the vegetative power in its cutting? (Sec. 233)

The case of anomalies in natural production had been explained by More as being due ultimately to the imperfections of matter and directly to the automatic quality of the Spirit of Nature. But Berkeley explains them as being directly intended by the Deity, suggesting that the natural world cannot be subjected to merely rational standards of right and wrong:

Natural productions, it is true, are not all equally perfect. But neither doth it suit with the order of things, the structure of the universe, or the ends of Providence, that they should be so. General rules ... are necessary to make the world intelligible, and from the constant observation of such rules natural evils will sometimes unavoidably ensue; things will be produced in a slow length of time, and arrive at different degrees of perfection (sec. 256).

It is somewhat contradictory that Berkeley should so easily pass over natural errors when his universal aether is constantly guided by Mind. But this is because Berkeley does not have any definite notion of the Soul as the intermediary between the Mind and the material world. Rather, 'Intellect' itself is

the very life of living things, the first principle and exemplar of all, from whence by different degrees are derived the inferior classes of life: first the rational, then the sensitive, after that the vegetative (Sec. 275).

By making the rational soul precede the sensitive and vegetative, Berkeley reverses the evidence of the sensible world, where vegetative and sensitive life always precede the rational. This false view of evolution is due to the fact that Berkeley has no place for Soul in his metaphysics and no concept of the dynamic relationship between Mind and matter which alone can show why spirit embodied in matter must necessarily start at the dimmest levels of life and then rise to enlightenment.

Although Berkeley vaguely refers to 'some Platonics' who held that nature is the 'act of the Soul' and the 'life of the world', he never achieves the clarity of More who, more Neoplatonically, sought to demonstrate this act of the Soul in his symbolic model of the universe, intelligible and sensible. In this model, we have seen, the Soul's energetic act is an 'audacious' spring towards the nothingness of matter so that the Spirit of Nature is merely the Soul reflected matter. Although Berkeley was opposed to the postulate of a Spirit of Nature, he adopts the same theories about the vital and organic processes of the universe and transfers them to the corporeal aether directed immediately by the Divine Mind. But in the absence of any concept of the Soul, the Mind is necessarily rendered defective. There is, thus, in Berkeley, virtually no metaphysical understanding of the imaginational and unconscious bases of the world. The extension that More had granted to spirit as well as to matter, on the other hand, allowed him to show how Mind and matter were equally extended, the latter imperfectly reflecting the former, so that Nature is basically an unconscious shadow of Reason. In fact, the Plotinian model of the universe as formulated by More reveals not only the way in which Mind motivates matter through Soul, which ranges between the intelligible and sensible realms as Divine energy, but also what Berkeley himself was at pains to establish throughout his career as a philosophical immaterialist, namely that Nature or the *phenomenal* world is but the final result of the Mind's positing, or *perceptio* of Itself and therefore must be as incorporeal as its original.

NOTES

[1] For a further discussion of some of the details of this aspect of More's philosophy see the Introduction and Notes to my edition of *The Immortality of the Soul*, Dordrecht: Martinus Nijhoff (1987). Also, R. A. Greene, 'Henry More and Robert Boyle on the

Spirit of Nature', *Journal of the History of Ideas* **XXIII**, pp. 451—474 (1962). For a more general survey of concepts of the plastic in the seventeenth century, see W. B. Hunter, 'The Seventeenth Century Doctrine of Plastic Nature', *Harvard Theological Review*, **43**, pp. 197—213 (1950).
[2] The literature on Plotinus is extensive. But for a good discussion of Plotinus' conception of Nature, see J. N. Deck, *Nature, Contemplation, and the One*, Toronto: University of Toronto Press (1967).
[3] All quotations from Plotinus are from *The Enneads*, translated by Stephen Mackenna, London: Faber and Faber (1969).
[4] The *Platonick Song of the Soul* is the final, expanded version of More's *Psychodia Platonica*, published in 1642.
[5] Recently, John Henry, in his article 'A Cambridge Platonist's materialism: Henry More and the concept of the Soul', *Journal of the Warburg and Courtauld Institutes*, **49**, pp. 172—195 (1986), has recognized the difficulty of considering More's position as a consistently dualistic one. But he erroneously considers More's non-dualism to be a materialistic one rather than an immaterialistic Neoplatonic one, as it was intended to be and, in fact, is. Henry fails to analyse the precise relationship between the definitions of spirit and matter in More and, basing himself on More's descriptions of created spirits or souls, which are always allied to some form of matter, concludes that More's system was a 'crypto-materialism'.
[6] The diagram is reproduced from 'Notes upon Psychozoia' in *The Complete Poems of Henry More*, ed. A. B. Grosart, Edinburgh (1878), rpt. Hildesheim: Georg Olms (1969), p. 148.
[7] This is More's English version of the Plotinian τόλμὰ.
[8] More, Notes upon 'Psychozoia', St. 41.
[9] More, *Ibid*. R. A. Greene who undertook a critical survey of More's use of the notion of the Spirit of Nature and the controversy regarding this entity with Robert Boyle is thus certainly wrong in finding a contradiction in More ('More seemed unaware that he could not logically make the Spirit of Nature consciously pursue a design activated by the good and at same time imperceptive and unintelligent') where there is indeed none. See R. A. Greene, *op. cit.*, p. 465.
[10] All quotations from *The Immortality of the Soul* are taken from Henry More, *A Collection of Several Philosophical Writings of Dr. Henry More*, London: printed by Joseph Downing (1712). The 1712 edition of his collected philosophical works is an English translation from his *Opera Omnia*, Londoni: Imprensis J. Martyn and Gualt Kettilby (1675—1679).
[11] Richard Burthogge, *An Essay upon Reason and the Nature of Spirits*, London (1694), p. 128.
[12] D. P. Walker, for example, in his article, 'Medical Spirits in Philosophy and Theology from Ficino to Newton', in *Music, Spirit, and Language in the Renaissance*, London (1985), pp. 287—300, refers to More's inclusion of the Spirit of Nature along with the World-Soul without understanding More's reason for distinguishing the two entities. Thinking that God, the Spirit of Nature, and the ether are, in More's system, 'all doing the same thing at the same time', Walker accuses More of presenting an 'unnecessary proliferation of cosmological principles'. But I have shown that it is precisely the subtlety of More's Neoplatonic system that demands these distinctions,

and only a failure to understand their significance could lead to false conclusions such as that More's world-view is 'a strangely corporealized version of a Neoplatonic hierarchy of emanations' (Walker, *op. cit.*, p. 295), or that it is a 'crypto-materialism' (Henry, *op. cit.*, p. 195).

[13] The Neoplatonic notion of the soul's three vehicles, ethereal, aerial, and terrestial — derived from Proclus' *Platonic Theology*, III, (v), 125f is explained by More in Book II, Chap. 14. He is careful to show that Aristotle too believed in an ethereal element in the seed, φύσις 'αναλογος οὖσα τῷ τῶν ἄστρων στοιχείῳ (*De Generatione Animalium*, II, iii, 736b). He even approximates this element to the first and second elements of Descartes (*Principia*, IV, 3—4). For a brief history of the Neoplatonic notion of the soul's vehicles, see E. R. Dodds's edition of Proclus' *Elements of Theology*, Oxford (1963), Appendix II 'The Astral Body in Neoplatonism', pp. 313—321.

[14] Cudworth, *The True Intellectual System of the Universe*, Preface, xxxviii. All quotations from Cudworth are from *The True Intellectual System of the Universe*, 2 Vols., ed. Thomas Birch, London (1743).

[15] *Op. cit.*, Ch. V, p. 644.

[16] *Ibid.*, pp. 769—70.

[17] This weakness is noted by Lydia Gysi, *Platonism and Cartesianism in the Philosophy of Ralph Cudworth*, Bern (1962), p. 24.

[18] 'And that more *Extensions* then one may be commensurate, at the same time, to the same Place, is plain, in that *Motion* is coextended with the Subject wherein it is, and both with *Space*. And *Motion* is not nothing; wherefore two things may be commensurate to one Space at once' (*The Immortality of the Soul*, Bk. I, Ch. 2, Sec. 12). In the *Divine Dialogues*, First Dialogue, Sec. 25, when Hylobares, the materialist remarks that 'Motion is not *Ens*, but *Modus Entis*.' Philotheus, representing More, replies that 'Motion is *Ens*, though in some sense [i.e. since it is the inner form of the body] it may be said to be *Modus Corporis*.' He later goes on to say that motion 'has an extension of its own, *reduplicative* into it self,' and that it is the '*vis agitans* that pervades the whole body that is moved.'

[19] John Passmore, *Ralph Cudworth: An Interpretation*, Cambridge (1951), p. 28.

[20] Cudworth, *op. cit.*, p. 885.

[21] *Ibid.*, pp. 593—4.

[22] All quotations from Berkeley's *Siris* are from *The Works of George Berkeley, Bishop of Cloyne*, Vol. 5, ed. T. E. Jessop, London: Thomas Nelson and Sons (1953).

[23] Newton, *Opticks* (1730 edition), N.Y.: Dover (1952), p. 354. For a lengthy discussion of Newton's theories of the aether, see J. E. Macguire, 'Force, Active Principles, and Newton's invisible realm', *Ambix* **15**, pp. 154—208 (1968).

JAMES FRANKLIN

THE ANCIENT LEGAL SOURCES OF SEVENTEENTH-CENTURY PROBABILITY

The Scientific Revolution of the seventeenth century might almost equally well be called the Mathematical Revolution. Experimentation was important, but even there one of the advances was the systemic extension of *measurement*, to such quantities as time, speed, pressure and probability. The initial reduction of data to numerical form enabled the discovery of patterns in them like the laws of motion, and the creation of mathematical theories of continuity (the calculus) and of probability.

In the case of the calculus, it is reasonably clear what exactly the subject matter, continuous motion, is. In the case of probability, it is notoriously otherwise, and disagreements about what is being studied have biased investigations of the history of the subject. There are two kinds of probability — *factual* or *stochastic* probability (dealing with the characteristic patternless sequences produced by stochastic or chance processes like the throwing of dice or coins), and *logical* probability (the relation of partial support or confirmation between one proposition and another, found in the relation of evidence to hypothesis, in 'proof beyond reasonable doubt', and so on). It is a matter of philosophical dispute whether one of these notions is reducible to the other, but in any case the surface distinction is clear, and provides an orientation in the history of the subject. However, there are three factors which confuse the issue further, and so make it more difficult to decide what to look for in early authors.

Firstly, actual people's partial belief in, or uncertainty or doubt about, propositions and the support they give one another (sometimes called 'subjective probability') has some relation to logical probability, but is not the same as it (indeed, psychological experiments show consistent discrepancies between the two).

Secondly, there are connections between logical and factual probability, in that if the outcome of an experiment has a factual probability other than 0 or 1, the experimenter is rationally uncertain of the outcome. Hence it will sometimes be unclear which kind of probability an author means, or whether he is aware of any distinction.

Thirdly, the *mathematical* theory of probability can be, and usually has been, developed without reference to what kind of probability is being spoken of. This is because the mathematical theory of, say, dice throwing depends only on the *symmetry* or *equiprobability* of certain outcomes, allowing equal numbers to be assigned to them. One can say that all of the 6 possible outcomes of a single throw of a die are symmetrical, and hence 'have probability 1/6', without it being necessary to say in what respect they are symmetrical, or what the probability consists in.

The resulting difficulties of interpretation (or, to be blunt, confusions) are found in the original writings of Pascal, Fermat, Huygens and the other founders of probability theory. They have, on the whole, become steadily worse in the mathematical and historical literature since.

THE BIRTH OF MATHEMATICAL PROBABILITY

The story of the beginning of mathematical probability is a familiar one. In 1654, Pascal wrote to Fermat concerning the problem proposed to him by the Chevalier de Méré, a gambler, of the correct division of the stake in an interrupted game of chance. Fermat and Pascal between them solved a number of particular problems (for example, What is the just division of the stake if two players have agreed that the first to win 3 points wins the game, and the game is interrupted when one has won 2 and the other 1?). Their method is essentially that of counting the equiprobable cases that comprise various outcomes, though they do not say so explicitly. Their letters became widely known and came to the attention of Huygens, who wrote a short book solving similar problems, published in 1657. Individual problems on dice were solved by many people, including Newton and Leibniz, and theorems of a more general nature were discovered, notably by James Bernoulli. Modern mathematical probability developed smoothly from these beginnings. Except in the case of Leibniz, there was almost no attempt to explain what probability was, or relate it to earlier developments.

In the 1670s, an entirely different mathematical theory was developed, that of calculating the just price of a life annuity. The correct answers require knowledge of the expectation of life for various ages. This must be found empirically, from examining records of mortality. It is not at all obvious that this has anything to do with the problems about dice. Nevertheless, the originators of the theory, Hudde, de Witt and Huygens, thought, rightly, there was a connection. The work on annuities was the ancestor of modern statistics, the science of making inferences from numerical data, but the science developed much more slowly than probability.

In trying to understand the relation of these developments to what went before, it is not clear where to look. If one looks for mathematics concerning dice, there is very little to find. Not much has been added to the results of F. N. David's *Games, Gods and Gambling* of 1962. In summary, the question, 'What is the just division of the stake in an interrupted game of chance?' had been mentioned briefly in Italian mathematical works from the fourteenth century, and had attracted a series of wrong answers.[1] In addition, Galileo, in an unpublished note, had solved two problems on dice by counting combinations; he gives the impression that such problems are easy and uninteresting (Fermat also sometimes seems to think the same). That is all, and there is every reason to think there is nothing of much significance left to discover. Two considerations arise. First, the problem of calculating the probability of outcomes at dice is, *prima facie*, indeed not very interesting. Secondly, even if it were, the question of the division of the stake in an interrupted game is a very unobvious place to start. There is even something paradoxical in thinking first of a *moral* question in connection with gaming.

It might, then, be useful to look not so much for 'anticipations' of what was new in the seventeenth century (that is, numerical quantification), but rather for what was being said about the phenomena that became quantified. (For comparison, the calculus in the seventeenth century achieved mathematical mastery of continuous motion, but this phenomenon itself has a long history of preparatory study in, for example, scholastic physics). As noticed in Lorraine Daston's recent study, *Classical Probability in the Enlightenment*,[2] a natural place to search is in law, that is, the continental continuation of Roman law. There are various reasons for the choice. The originators of mathematical probability were all either professional lawyers (Fermat, Huygens, de Witt and Leibniz) or at least the sons of lawyers (Cardano, Pascal and Arnauld), so can be presumed to be familiar with at least the broad concepts of legal thought. The two questions which mathematical probability answered were both problems in the law of contract (namely, what is the just division of the stake in an interrupted game?, and, What is the just price of a life annuity?). Again, law was the most developed of the social sciences before modern times, and of its nature is forced to consider carefully and explicitly matters of uncertainty (such as doubtful evidence) and chance phenomena (such as mortality in connection with annuities and insurance).

The consideration of the two kinds of probability is unconnected in the legal tradition, so it is convenient to study them separately.

CHANCE PHENOMENA IN ROMAN LAW

The classical period of Roman law is that of the jurists Ulpian, Papinian and others of the period 150—250 A.D. Excerpts from their works were collected, and sometimes adjusted, by a commission set up by the Emperor Justinian, and headed by Tribonian. This collection, the *Digest* or *Pandects*, was declared law in 533; all other writings of the jurists were declared no longer law, and have almost completely disappeared. The *Digest* forms the largest part of Justinian's complete codification of Roman law, the *Corpus of Civil Law*. The details of what is the fifty books of the *Digest* are important for the history of thought, since every word in it was minutely examined, compared with others, and commented on for centuries. The centuries in question — roughly the twelfth to the seventeenth — were ones in which law occupied a larger place in intellectual life than it does today, and in which thought was, in any case, less compartmentalised.

There is in the *Digest* what may or may not be a table of life expectancies. The problem being considered is how to decide on a reasonable figure for the present value of a life annuity, that is, an annual payment that will continue for a person's lifetime:

Ulpian says that the following rule should be adopted in making the estimate of maintenance to be furnished. The amount bequeathed to anyone for this purpose from the first to the twentieth year is computed to have lasted for thirty years ... From twenty to twenty-five years the amount is calculated for twenty-eight years, from twenty-five to thirty years the amount is calculated for twenty-five years; from thirty to thirty-five years, the amount is calculated for twenty-two years, from thirty-five to forty years, it is computed for twenty years; from forty to fifty years, the computation is made for as many years as the party lacks of the sixtieth year after having omitted one year; from the fiftieth to the fifty-fifth year, the amount is calculated for nine years; from the fifty-fifth to the sixtieth year, it is calculated for seven years; and for any age above sixty, no matter what it may be, the computation is made for five years.[3]

It seems unlikely that Ulpian used any data of mortality to arrive at these figures,[4] but the diminishing of the value of an annuity with age does seem designed to reflect very roughly the diminishing of life expectation.

The *Digest* has something to say about contracts whose fulfilment depends on chance:

Neither a purchase nor a sale can be held to take place without a thing that can be sold ... A sale is, however, sometimes understood to be contracted without a thing being the object of the same as, for instance, where a purchase is made dependent on chance (*quasi alea*), which occurs when a catch of fish or birds, or money to be thrown to the populace, is bought; a purchase is contracted even if nothing happens, because it is the hope that is sold.[5]

The opposite of a 'hope' is a 'peril' or 'risk'. From the time of the Roman Republic, shipwreck was 'at the risk of the state' (leading to some predictable frauds).[6] The point in various kinds of sale at which the risk passed from the seller to the buyer was laid down,[7] and the bearing of such a risk could entitle one to benefits.[8] Rates of interest were controlled by law; for money carried by sea, 'maritime interest' at a higher rate was allowed, because of the assumption of the 'uncertain peril' of the voyage.[9] The *Digest* says, 'The price is for the peril'.[10] There is a major conceptual advance in thinking of a 'hope' and a 'peril' as quantities that can have a price.

It is important to consider medieval developments, because the seventeenth century understood Roman law as developed by the medieval jurists. There was much less of a Renaissance rejection of the medieval heritage than in other fields.

Early capitalism, perhaps even more than any other period, was a time when ready money was necessary, but hard to raise. One tempting source, for states and other powerful bodies, was the sale of something in the future, or the prospect of it, for hard cash in the present. In the thirteenth and fourteenth centuries, there grew up a trade in both perpetual and life annuities (in which one pays a sum of money and receives a stated annual income, in perpetuity or for life respectively). There is an element of betting involved in life annuities, in the seller's estimate of the likely date of the purchaser's death. In 1308, the Archbishop of Bremen, on payment of 2400 livres to the Abbey of St Denis, acquired the right to an annuity of 400 livres. There was provision for a partial refund in the event of his death within two years, and presumably the Abbey was counting on paying out money for not too great a multiple of that period. The Abbot contested the validity of the contract, unsuccessfully, on the ground that it was usurious and therefore not binding, in 1323, and the Archbishop lived until 1327.[11] The cities of Germany commonly raised money through the sale of annuities from the thirteenth century. Hamburg standardly offered 10% interest on life annuities, as against 6.66% for perpetual annuities; the difference represents some kind of implicit quantification of the expected life span.[12] The risk involved in annuities generally was part of the justification for exempting them from the usury laws.[13] In the fourteenth century, there is found the first discrimination on grounds of age, to reflect the different expectancies of life.[14]

Cases such as the Archbishop's brought annuities to the attention of lawyers. In the ideas of the *Digest* on the prices of hopes and perils, the lawyers had, in theory, a conceptual tool for analysing the matter, but it would seem too much to expect that they would understand this, since it involves understanding the probabilistic aspect of life annuities, as opposed to perpetual ones. But in fact they did understand this perfectly well. The *Consilia* of Baldus, of the later fourteenth century, discuss cases in which someone buys a life annuity for 300 florins, and survives long enough to receive a total of 600. Is this usury?

It would seem not, because a life annuity can be bought with a good conscience (*Code* 1.2.14). Again, the contract is not usurious in form, nor should it be considered usurious in intention; for what if the buyer lived no more than one day? And so the contract seems like the buying of fortune (*Dig.* 19.1.12 on the cast of a net).[15]

Again, the contract is not usurious; for usury is the increase of a returnable stake, but here the stake is not to be returned; therefore nothing accrues to the stake, so it is not a usurious contract ... Again, the buyer subjects himself to the peril of fortune, whence the price is of the peril (*Dig.* 22.2.5 on the price of peril). Again, if the buyer had lived only one hour, the seller would not have given the price back to him: certainly not, whence one ought not to demand from someone else what he would not take into account himself, or the contrary of which would not be demanded. Again, here a hope seems to be sold, so whatever the hope is worth is to be considered, and this is not to be judged from the outcome (as per Bartolus on *Dig.* 39.4.11.5). Again, it seems age is to be considered (*Dig.* 35.2.68), whence is taken conjecture of life and death. Again, an old man is dealt with differently from a young man (*Dig.* 2.15.8.10), on account of age. Again, by reason of the uncertainty the mean is said not to be exceeded (*Code* 4.32.17) ...

because a contract where there is equality of potential loss and gain stands far distant from the evil of usury, or from the attraction of usury, because of the doubtfulness of the peril (*Dig.* 17.2.29.1) . . . Notwithstanding the [*Decretals*] cap. *Naviganti* and cap. *In civitate*, where fraud is presumed: because there even if there was not the certainty of necessity, it was however very close to certainty, but here it is not, so the case is different.[16]

As to the actual pricing of life annuities, Baldus says:

So it is sold at some suitable and just price, namely, whatever such a hope can be commonly sold for at the time of the contract. And a hope is not worth as much as a thing, since a hope may be interrupted by many chance events: just as a crop is worth less on the stalk than in the barn.[17]

Baldus remained one of the most widely read legal authors for several centuries, and his opinions can be regarded as common knowledge in the legal profession. His views on annuities in particular were discussed and expanded in 1546 by Du Moulin, who considered how much a life annuity should be worth, compared to a perpetual one.[18]

Insurance was an Italian invention of the fourteenth century, the first contracts dating from the 1340s. Being a contract, an insurance is essentially a legal entity, and one that involves a fairly explicit quantification of risk. The early contracts do not exactly say this, but, like everything else in commercial practice, the legitimacy of insurance soon became debated by jurists and theologians who did have the background to explain the nature of the contract. The conceptualisation of 'risk' involved was based on that in Roman law, (the first book on insurance was a legal treatise[19]) but made a certain advance towards explicit quantification. Domingo Soto, confessor to the Emperor Charles V, was one of the first to write a book *De Iustitia et Iure*, a common title in the next hundred years. In considering whether insurance is legitimate, he does not take the question very seriously, thinking that of course it is legitimate, since it is necessary for business. He notes possible difficulties with the bull *Naviganti* and the view of some that insurance involves paying for something which does not exist, a 'peril', but replies:

For anything that can be estimated at a price, one can receive a fee: to render a thing safe [insure it], which is exposed to peril, can be estimated at a price . . . we say of a fair game: whether it will rain tomorrow or not, etc; so in the same way it is permitted to expose a thousand ducats, say, to peril with the hope of making fifty or sixty. There are some who regard it as stupid to allow the peril of someone's ship worth perhaps twenty or thirty thousand, in the hope of making a hundred or a thousand. To this we reply that it is not for us to dispute about prices: these can be just or unjust, but it is for the contracting parties to decide them. But there is no stupidity or folly in accepting this kind of peril at the going price; in fact nothing is more obvious than that insurances can expect to gain. They may lose sometimes, but at other times they accumulate gain.[20]

Life annuities remained standard methods for raising public finance. The city authorities of Amsterdam sold a life annuity in the year 1586—90; the records were later the basis for the calculations of Huygens, Hudde and de Witt.[21]

Such questions were considered in moral theology as part of the justice of contracts, under the subheading of contracts involving chance. Discussions on fair division in cases of uncertainty can be found in the works of such casuists as Lessius and Bauny, two of the authors later criticised in Pascal's *Provincial Letters*. Lessius' *De Iustitia et Iure* of 1605 considers the case of an investor who contributes 1000 gold pieces to a joint venture, and is solely liable for the loss of this sum in case of misfortune. In determining the just partition of the profits, Lessius estimates that the 'peril' is worth 100 gold pieces. To the objection that it is not 100 but 1000 gold pieces that the investor stands to lose, Lessius replies, 'The uncertain peril of chance should be reduced to a certain price', and imagines a fictitious contract of insurance that the investor could enter into, paying a premium of 100 gold pieces for the safety of the 1000.[22] Still closer to Pascal's time is the explicit quantifying of 'hopes' by Bauny:

What return should be offered to those who take upon themselves the perils and other chance events to which everything is subject in commerce, and in particular, what is the money which is the sum proportionate to the indefinite and uncertain gain one can expect from 100 écus invested in a Merchant Company? One will consult one's conscience, and experts; generally, it can be said that it should equal the hope one has of drawing more or less profit, 'which is able to value the uncertain gain hoped for, by the certain gain of many, this gain being diminished according to the size of the peril', as Major says in q. 49 . . .[23]

The crucial notion of a 'price equal to a hope', that is to a numerical probability, is common among the casuists. Thus in his chapter on 'gaming, wagers and insurance', the Jesuit cardinal, Juan de Lugo says:

Therefore the first condition for the justice of an insurance is, that the price be equal to the peril undertaken; certainly that the price paid for the obligation should be as much as that obligation is worth in the judgement of experts. This price is not a definite amount, but has a maximum, mean and minimum, as with buying and selling. As varied circumstances affect the peril, so the just price should be varied. The equality is to be taken from the quantity of the peril at the time of the contract, not after the event.[24]

Lugo goes on to consider the morality of cases where the 'peril' is unequal for the two parties, as when one knows the ship is safe, or lost, or knows from 'astrology' that there will be no storms, or from private communications that there are no pirates.

This brings us to the Fermat-Pascal correspondence. While the first letter, from Pascal, is unfortunately missing, how the two conceived of the question can be gathered from some remarks in the first letter from Fermat:

Monsieur,
If I undertake to make a point with a single die in eight throws, and if we agree after the money is put at stake, that I shall not cast the first throw, it is necessary by my theory that I shall take 1/6 of the total sum to be impartial because of the aforesaid first throw...
But you proposed in the last example in your letter (I quote your very terms) that if I undertake to find the six in eight throws and if I have thrown three times without getting it, and if my opponent proposes that I should not play the fourth time, and if he wishes me to be justly treated, it is proper that I have 125/1296 of the entire sum of our wagers.[25]

The rest of the correspondence is mostly mathematical; the phrases, 'the lot is such that . . .', 'the hazard is equal', 'a just division', and 'without disadvantage or advantage' are all that can guide us.

Huygens, then aged 26 and little known, visited Paris in 1655 (on his way to the University of Angers to purchase a doctorate in law). He learned of Fermat and Pascal's achievements, but did not visit Pascal, believing him to have withdrawn into purely religious concerns. On returning to Holland, Huygens wrote, in Dutch, a small treatise, *On Reckoning in Games of Chance*. A Latin translation was published in 1657, attached to van Schooten's *Mathematical Exercises*, and the Dutch original in 1660.[26] It is the first published work on mathematical probability. It introduces the subject matter thus:

Although in games determined solely by chance the outcomes are uncertain, there is always a fixed value for how much one has for winning over losing. Thus if someone bets on throwing 6 points on the first throw of a die, it is uncertain if he will win or lose, but how much more likely (*quanto verisimilius*) he is of winning his bet than of losing it is something determinate that can be subject to calculation. Similarly, if I play against someone with the winner the first to win three, and I have already won one, it is uncertain which of the two will win. But how much my expectation (*expectatio*) is, and how much his, can be estimated exactly by calculation, and one knows consequently how much greater, if we interrupt the game, the part of the stake rightly mine exceeds his. One can calculate also at what price it would be fair (*aequum*) to sell my place to someone who wished to continue the game in my place . . .

I take it as a foundation that in a game the chance (*sortem seu expectatio*; in the Dutch simply *kansse*) that one has towards something is to be estimated as such that, if one had it, one could procure the same chance in an equitable game (*aequo conditione certans*; the Dutch adds: that is, a game where no loss is offered to anyone). For example, if someone hides from me 3 shillings in one hand and 7 in the other, and gives me the choice of taking either hand, I say this is worth the same to me as if I were given 5 shillings. For if I have 5 shillings, I can again arrive at having an equal chance (*aequam expectationem*) of getting 3 or 7 shillings, and that by an equitable (*aequo*) game.

Proposition 1. If I can get a or b, of which either is equally easy for me, my expectation should be said to be worth $(a + b)/2$.[27]

Further propositions evaluate what various combinations of chances are worth; the further content is mathematical. The device used is the same as in Lessius, the estimation of the monetary value of a chance by a fictitious sale.

The problem treated by Jan de Witt in 1671 was that of deciding how much life annuities were worth, compared to perpetual annuities. To decide the worth of an annuity, he uses a hypothetical mortality table, obtained from smoothing an actual table by using the symmetry principle that the chance of death in two consecutive half-years should be almost equal. He estimates that it is reasonable to take the chance of death in any half-year as equal for ages 3 to 53, then 3/2 times this for ages 53 to 63. To explain how this bears on the value of annuities, de Witt, like Pascal and Huygens, finds it natural to express himself in terms of a fair contract: a fair contract can be made between a man of 40 years and a man of 58 as follows: if the man of 58 dies within 6 months, the man of 40 inherits 2000 florins, but if the man of 40 dies within 6 months, the older man inherits 3000.[28]

THE PROBABILITY OF EVIDENCE

The discussions of chance phenomena do not use the words 'probable' (*probabilis*) or 'likely' (*verisimilis*). Those words are reserved for a much more widespread theory, concerning belief in propositions for which certainty cannot be attained. It is found principally in law, but extends into related subjects such as moral theology, rhetoric and logic, and is found occasionally in the evaluation of scientific theories (especially in Galileo and Kepler) and in religious controversy. It has been argued, in Hacking's *Emergence of Probability*, that this tradition is not relevant to mathematical probability, but was if anything a hindrance to it, on the grounds that a 'probable' opinion was not one supported by evidence, but one approved by some authority, or by the testimony of respected judges.[29] This claim is not true, as we shall see below; probability as used in law *did* mean 'well supported by evidence', though of course some of that evidence might be the testimony of witnesses. Leibniz, the only one of the protagonists to say much about what probability was conceived to be, was in no doubt that the legal theory was connected with the mathematical one. His *New Essays* of 1704 contain a brief description of the theory:

Jurisconsults in treating the proofs, presumptions, conjectures and indices have said a number of good things on this subject (of degrees of assent), and have entered into some considerable detail. They begin with *notoriety* ... There are *proofs more than half complete* ... *conjecture* and *indices* ... the entire *form of juridical procedures* is in fact nothing but a species of Logic applied to questions of law. Physicians also have a number of degrees and differences in their *signs* and *indications* which may be seen among them.
The Mathematicians of our day have begun to estimate chances in connection with gambling games...[30]

Again, one must start with the *Digest*, since everything later constitutes a use of its concepts. The fate of the *Digest* contrasts with that of some earlier works which made a contribution to probability. Reasonable ideas on the evaluation of inconclusive evidence appear in some of the Attic orators, Aristotle's *Rhetoric*, and some ancient medical writers, but these were little read until modern times. Cicero and the *Rhetoric to Herennius* give some attention to the 'probable' (*probabilis*) and the likely (*verisimilis*), but the concepts are ill-defined, even almost vacuous.

The concept of the probable expressed by Cicero in the synonyms *probabilis* and *verisimilis* begins to do active work in the *Digest*. In fact, it is the first place where there is an easy use of the day-to-day concept of 'probable'. Many of the instances involve making a reasonable estimate of someone's intention, for example, the intention of someone who has made a will and is now dead:

I answered that it is very likely (*verosimilius*) that the testator had intended rather to point out to his heirs where they could readily obtain forty aurei ... than to have inserted a condition in a trust.[31]

It is not likely (*verisimile*) that she intended the benefit to be enjoyed by the substitutes sooner than it could have been by her son.[32]

A person who was already exercising the functions of a public office afterwards undertook the construction of an aqueduct. It seemed to be absurd for him to ask to be released from his former employment, when he was already charged with both; because if he had only intended to assume responsibility for one, it is more probable (*probabilius*) that he would have obtained exemption from the other.[33]

That these words really do mean 'probable' is confirmed by the fact that such arguments can be overcome by proofs to the contrary:

It is likely (*verisimile*), however, that in this instance also, the party who gave the dowry had a view to his own interest; for he who made the gift on account of the marriage can, if the marriage is not performed, bring an action for recovery as if on the ground of want of consideration, unless the woman should be able to show by the most evident proofs that he did this rather for her benefit than for his own advantage.[34]

There is one instance where the uncertainty is about an event rather than an intention:

The woman lost her life by shipwreck at the same time as her child, who was one year old. For the reason that it appears likely (*verisimile*) that the child died before its mother, it was decided that the husband should retain a portion of the dowry.[35]

An explicit link appeared between probability and 'presumptions'. Sometimes presumptions were, as in ancient Jewish law and the modern 'presumption of innocence', methods of reaching decisions under uncertainty, irrespective of whether the decision was probable or not. The Carbonian Edict declared that if there was a doubt as to whether a child was in fact the issue of someone who had left money in a will, the presumption was in favour of the child.[36] That is, the question was settled for moral reasons, without considering how probable the outcome was. But at other times, a presumption clearly does mean what is probable:

It is not likely (*verisimile*) that a person would pay in a city, under compulsion and unjustly, something which he did not owe, if he showed that he was of illustrious rank; since he could invoke the public law, and apply to someone in authority who would forbid his being treated with violence. The most open proofs of violence must be given to oppose a presumption of this kind.[37]

There can be grades of presumption:

The same rule must be said to apply if he believed with a not light presumption that the property which was purchased had come into his hands as a part of his father's estate.[38]

It must be emphasised that Roman law is very untheoretical as compared to modern law. While 'probability' and 'presumptions' are used in reasoning about particular cases, there is no account of what probability or presumptions are, or any general rules about them.

The medieval west was the heir of the two law-soaked ancient civilisations; even before the discovery of the *Digest* the West's sacred book was one divided into two 'testaments', and full of laws, trials and witnesses. From the Old Testament came the two-witness rule: Two witnesses are required for the conviction of anyone on a criminal charge.[39] The Glossators, unlike the Roman jurists themselves, were concerned with matters of general principle. They sought to harmonise the text by finding distinctions based on matters of principle, often of extreme generality. The most significant development for probabilistic argument was the Glossators' invention of the category of 'half-proof' (*semiplena probatio*). It was a natural thought that, if two witnesses are, in theory, full proof, then one witness is half-proof. The Glossator Azo says, around 1200:

It would seem this does not hold, because either the plaintiff proves, or not. If he proves, the defendant should be condemned. If not, he is acquitted, according to the rule ... I reply that although according to Aristotle it would seem to be an exhaustive division, it is not so according to the laws. There is a medium, namely, half-proof. Say therefore that in such a case proof has been less than full. Then the judge gives the right to complete the case by oath to the plaintiff. Otherwise if he has not proved at all, the defendant is acquitted, and an oath is not allowed, except in cases where the law gives the defendant the right to clear himself by oath ... The right to an oath is therefore given to the plaintiff if he half proves, either by private documents, or by the flight of the defendant who ... or by one witness ... Full proof is by two witnesses, therefore half proof by one.[40]

Canon law developed in parallel with civil law, an authoritative text being created out of the decrees of Popes and Councils, as collected in Gratian's *Decretum* of about 1150 and the *Decretals* of 1234. Innocent III, a canon lawyer before becoming Pope, was particularly concerned with questions of evidence. In a decree of 1209, he considers the advice a priest should give to one who doubts the validity of his marriage because he suspects some impediment, such as consanguinity:

> We believe it should be distinguished, whether the spouse knows for certain the impediment to the marriage, and then he may not engage in carnal intercourse without mortal sin, even though he could not prove it [the impediment] before the Church; or whether he does not know the impediment for certain, but only believes it ... In the second case, we distinguish, whether his conscience is thus from a light and rash belief, or a probable and discreet one (*ex credulitate levi et temeraria, an probabili et discreta*) ... when his conscience presses his mind with a probable and discreet belief, but not an evident and manifest one, he may render the marriage debt, but ought not to demand it.[41]

The importance of this text is the statement in the abstract that there are degrees of rational belief: the 'probable and discreet' is the grade *between* the 'light and rash' and the certain. The division is close to the Glossators' divisions of presumptions and proofs, and in fact it is not clear whether this threefold division appeared first in civil or in canon law. Its source seems to be a creative interpretation of a brief remark by the Emperor Hadrian in the *Digest*. There it had simply said:

> Slaves are to be subjected to torture only when the accused is suspected and proof is so far obtained by other evidence that the confession of the slaves alone seems to be lacking.[42]

The grades of suspicion arise by separating the clauses of this sentence. As a later slightly writer on canon law explains it,

> Proof has its grades ... This is proved by [The *Digest* title] 'On Tortures' 1.1 there: 'when the accused is suspected', that is the first grade; 'by other evidence', that is the second, 'that the confession of the slaves alone seems to be lacking', that is the third.[43]

These ingenious jurists have found a theory of grades of proof in the *Digest*, but plainly they had a good idea of what they wanted to find before they went looking for it.

That these grades were taken to be properties of the evidence, rather than just a subjective matter of someone's opinion, is confirmed by Aquinas, who repeated these distinctions some 40 years after Innocent, with examples added:

> Suspicion is of three kinds: One is the violent, to the contrary of which proof is not admitted, as when someone is found alone with a woman, naked on the bed, in a secret place, at a time apt for intercourse. The second is the probable (*probabilis*) as when

they are seen talking alone together in suspicious places. Third is the rash, which arises from a light conjecture ... although the first kind does not have the certainty of something actually sensed, or the certainty of demonstration, it does have certainty sufficient for proof in law. For the same kind of certainty is not required in all things, as is said in [Aristotle's] *Ethics* bk 1 ch 2.[44]

Aquinas mentions the half-proof of the lawyers, but without special comment.[45]

Medieval legal ideas percolated into the wider civilisation through a number of channels, mostly related to the Church's commitment to enforcing a legally based morality on the general populace. The 1215 decree of Innocent III and the Fourth Lateran Council required that every adult in Christendom attend confession once a year. It is no accident that the common name for the sacrament of penance is the same as that of the 'queen of proofs' in law; the confessional was regarded as a kind of miniature court of canon law. *Summae* for confessors treated individual morality in legal ways, distinguishing the various grades of 'doubts', 'scruples', 'moral certainty' and so on that the penitent might have, and the appropriate action in each case.[46] It was in this context that the doctrine of 'probabilism' eventually developed in moral theology. (The doctrine, first stated by Bartholome de Medina in 1577, holds that one may follow a 'probable' opinion in morality, even if the opposite opinion is more probable; an opinion is 'probable' if there are respected doctors who hold it and sound reasons in its favour).[47]

Despite humanist attacks, the elaborate medieval developments in continental law remained generally intact, though without notable development. Roman law was 'received' in Germany, replacing local customary law. The code known as the Carolina, of 1532, was intended as a summary of all the new criminal law in German, usable by officials without special training in law. Nevertheless, the law of evidence in it contains all the medieval details of half-proof, indices and presumptions. As usual, half-proof, such as one witness or an article owned by the suspect found at the scene of the crime, is sufficient for torture. Less than half-proof are 'sufficient indications', such as bad reputation, previous criminal record, flight, or enmity towards the victim. One indication is insufficient for torture, but two are sufficient if the judge thinks that together they amount to half-proof.[48]

At the level of academic law, Menochio's *Commentary on Presumptions, Conjectures, Signs and Indications* ran to two volumes,[49] and full treatments are found in Alciati's *Treatise on Presumptions*[50] and Mascardi's *On Proofs*.[51] These are the works containing the full version of the legal theory of evidence available to Leibniz. (The English law of evidence remained comparatively undeveloped and untheoretical, but in the early seventeenth century Coke, in particular, took over from continental law a considerable part of the apparatus of grades of presumptions.)[52]

In the century which included the scientific revolution and the peak of Dutch fortunes, the chief ornament of Holland in the social sciences was Hugo Grotius. His work was centred on, though far from confined to, law. His chief work, *The Law of War and Peace*, is regarded as the classic of international law; its legal topic is probably responsible for its status as the last influential work of fully scholastic cast. He took a close interest in the writings of the Jesuits, and appears to have studied Lessius' *De Iustitia et Iure* thoroughly.[53]

The *Law of War and Peace* of 1625 is a descendant of the works of the Spanish moral theologians Vitoria and Suarez on the same subject, and on the matter of doubts about the justice of war, his treatment is much the same. He does, however, take a somewhat stricter view of what is required in cases of doubt:

It seems then that the above-mentioned view of Adrian is to be followed, if the subject not only doubts, but, led by more probable arguments, leans more towards thinking that the war is unjust; especially if it involves attacking others, not defending one's own.[54]

('Adrian' is Pope Adrian VI, who had maintained that if a soldier doubts the justice of the Prince's war, he is obliged not to serve in it.[55]) Grotius' treatment here is in fact only a summary of that in a youthful work, *On the Law of Booty*. Written in 1605 in reply to attacks on the Dutch East India Company's policy of piracy on the high seas, the work remained unpublished, apparently because of the vanishing of criticism on distribution of the Company's profits.[56] There he had written:

Again, very few facts are discernible through the senses, since we cannot be in more than one place at one particular time, and since the senses perceive only those things which are very close at hand. Yet there is no other way of attaining to true knowledge. Impelled thus by necessity, human reason has fashioned for itself certain rules of probability, or *ton eikoton*, for passing judgement in regard to facts. These rules consist of various *prolepseis*, or (to use the Latin term) *presumtiones*, which are not fixed and unchangeable like scientific rules but rather of a character considered concordant in the greatest possible degree with nature; that is to say, on the basis of what commonly occurs, conclusions of a similar trend may be drawn. For, among the proofs which we accept in forming judgements, there is not one that is necessarily conclusive; on the contrary, all of them are derived from the aforesaid presumptions *hos epi to poly*, from what happens for the most part.

[he refers to Aristotle, the *Digest*, the *Decretals* on presumptions and Aquinas]

... And when the magistrates hold that things justifying entry into war have befallen the citizens, why should not faith be placed in those authorities...

For subjects, that war has a just cause which is ordered by a superior, provided that the reason of the subjects is not opposed thereto with probability.[57]

THE FUSION OF THE TWO PROBABILITIES

The first connection noticed between dice and the evaluation of evidence, and the first use of the word 'probability' in connection with gambling, occurs in Chillingworth's *Religion of Protestants a Safe Way to Salvation*, of 1638. He attacks the Catholic demand for certainty in matters of religion:

... you take occasion to ask, 'shall I hazard my soul on probabilities, or even wagers?' As if whatsoever is but probable, though in the highest degree of probability, were as likely to be false as true! Or, because it is but morally, not mathematically, certain, that there was such a woman as Queen Elizabeth, such a man as Henry VIII, that is, in the highest degree probable, therefore it were an even wager there were none such! By this reason, seeing the truth of your whole religion depends finally upon prudential motives, which you do but pretend to be very credible, it will be an even wager that your religion is false. And, by the same reason, or rather infinitely greater, seeing it is impossible for any man (according to the grounds of your religion) to know himself, much less another, to be a true pope, or a true priest; nay, to have a moral certainty of it; because these things are obnoxious to innumerable secret and undiscernible nullities, it will be an even wager, nay, (if we proportion things differently), a hundred to one, that every consecration and absolution of yours is void, and that whensoever you adore the host, you and your assistants commit idolatry: that there is a nullity in any decree that a pope shall make, or any decree of an council which he shall confirm: particularly, it will be at least an even wager, that all the decrees of the council of Trent are void, because it is at most but very probable that the pope which confirmed them was the true pope. If you mislike these inferences, then confess you have injured Dr. Potter in this also, that you have confounded, and made all one, probabilities, and even wagers. Whereas any ordinary gamester can inform you, that though it be a thousand to one that such a thing will happen, yet it is not sure, but very probable.[58]

Chillingworth is at ease with giving at least approximate numerical odds to chances with dice.

The next connection between gambling and belief appears in the famous wager of Pascal's *Pensées* (though the essential idea is to be found in Chillingworth.[59]) Strictly speaking, however, the wager as Pascal presents it has nothing to do with belief. Pascal uses the analogy of a game of chance for the situation of someone deciding whether to *act* religiously (by going to Mass and so on). Pascal suggests that, since the payoff is infinite if religion is true, it is worth one's while accepting the bet and acting religiously, *whatever* the evidence for the truth of faith. It is understood that going to Mass will eventually lead to belief, but the decision arising from the wager is about action, not belief. This is signalled by the non-occurence of the word 'probability' in Pascal's treatment.[60]

The actual fusion of quantification with logical probability occurs in the final chapter of Arnauld and Nicole's widely influential *Port Royal Logic*, of 1662, where numbers are first applied to something called 'probability'. Quantification is introduced, naturally, through games, and we find the expression that loss is 'nine times more probable than gain'. The concept is then applied to the old legal problem of the weight to be attached to a legal document properly drawn up by a notary. Since early medieval times, a public document had been taken as inducing a 'violent presumption' that the facts were as stated in it.[61] The *Port Royal Logic* explains this by saying that 999 out of 1000 duly notarised contracts have been properly dated (for example), so 'it is incomparably more probable that the contract before me is one of the 999 rather than the single one of the 1000 that is post-dated'.[62] (This is not to say, of course, that quantification here is a good thing, especially as the figures are obviously fictitious.)

Leibniz represents even more the fusion of old legal questions with the new efforts at quantification. Leibniz, to begin with, knew everything — both law and the studies of Pascal, Fermat, Huygens and de Witt. His interest in the subject originated with his early dissertation on conditional rights in law, of 1665.[63] His legal training was of course based on the ancient texts; his remarks on logic in law are illuminated by his 'paradox which, though amusing, gets at the truth: there are no authors whose style is more akin to the geometers than the old Roman jurists in the *Pandects*'[64] Conversely, he spoke of the theory of probability as a 'natural jurisprudence'.[65] Leibniz's actual attempt to seriously combine the legal concepts with the new quantification occurs in his unpublished manuscript of 1678, *De incerti aestimatione*. He tries to explain more clearly than Huygens what makes a game just:

A game is just if there is the same reason for hope and fear on each side. In a just game a hope is sold for what it is worth, because it is just to sell something for what it is worth, and the price of the hope is as much as the fear.

Axiom: If players deal similarly so that no difference can be assigned between them, except what consists in the outcome, there is the same reason for hope and fear.[66]

The point of this axiom is to allow symmetry arguments and thus introduce the mathematics:

If several outcomes are equally easy, and in some outcomes I will have the thing, and in the others lack it, the estimation of the hope will be the proportion the thing has to the whole, which is as the number of outcomes which are favourable to the number of all outcomes.

This is the first appearance of the modern way of thinking of probability as a ratio of 'favourable' outcomes to all outcomes. The phrase 'estimation of the hope' takes us back to the origin of the whole project in the *Digest*.

Leibniz represents, however, not only the coming together of legal and mathematical probability, but also their falling apart. There was an old legal saying (which Leibniz knew[67]). 'reasons are not to be counted, but weighed'. The fundamental problem in trying to apply mathematical probability to evidence in law (or, for that matter, in science) is that there seems to be no set of equiprobable basic alternatives. The end of the project may be taken to be the correspondence of 1703 between Leibniz and James Bernoulli. Bernoulli asks Leibniz for legal examples to which one could apply the determination of probabilities *a posteriori* (i.e. determining probabilities by experience, as in estimating the likelihood of death from mortality tables). Leibniz is unable to supply anything; he replies that such calculations are not usually necessary, as one can simply enumerate cases.[68] But since he had never produced any real legal cases solvable by such *a priori* quantification either, even to his own satisfaction, he seems to be making an empty claim.

Law had then served its purpose for the mathematical theory of probability. The service was never returned. Legal probability has continued to exist, and it is accepted in legal theory that such notions as 'proof beyond reasonable doubt' involve 'probability'. But all attempts to quantify the concept have been resisted.[69] Mathematical probability, meanwhile, has gone its own way.

NOTES

[1] Toti Rigatelli, L., 'Il 'problema delle parti' in manoscritti del XIV e XV secolo', in M. Folkerts and U. Lindgren (eds), *Mathemata: Festschrift für Helmuth Gericke*, Stuttgart (1985), pp. 229–236; David, F. N., *Games, Gods and Gambling*, London (1962), chs. 4, 6; Coumet, E., 'Le problème des partis avant Pascal', *Archives internationales d'histoire des sciences* **17**, 244–272 (1965).

[2] Daston, L., *Classical Probability in the Enlightenment*, Princeton (1988), ch. 1.

[3] *Corpus Iuris Civilis, Digest* 35.2.68.

[4] Greenwood, M., 'A statistical mare's nest?' *Journal of the Royal Statistical Society* **103**, pp. 246–8 (1940).

[5] *Dig.* 18.1.8; cf. 18.4.7; see Guitton, H., 'Le droit romain en face de l'aléa', *Revue Française de Recherche Opérationelle* **7**, pp. 194–5 (1963).

[6] Livy, *History*, 23.49; 25.3; Suetonius, *Claudius*, 18; *Theodosian Code*, 13.9.1–2.

[7] *Dig.* 18.1.35.7; 18.6.1; 19.2.13.5; *Code* 4.48.2.

[8] *Corpus Iuris Civilis, Institutes* 3.23.3.

[9] *Corpus Iuris Civilis, Code* 4.33.2; *Dig.* 22.2.4.

[10] *Dig.* 22.2.5; cf. Huvelin, P., *Etudes d'histoire du droit commercial romain*, Paris (1929), pp. 98–110.

[11] Jack, A. F., *An Introduction to the History of Life Assurance*, London (1912), p. 172.

[12] Baum, H.-P., 'Annuities in later medieval Hanse towns', *Business History Review* **59**, 24–48, at 28–9 (1985).

[13] P. 30.

[14] Jack, p. 175; *The Cambridge Economic History of Europe* (ed.) M. M. Postan, Vol. 3, Cambridge (1963), pp. 531–2.

[15] Baldus, *Consilia*, Vol. 2, cons. 154, Venice (1575), Vol. 2, p. 81.

[16] Vol. 3, cons. 210 (Vol. 3, p. 121).

[17] Vol. 5, cons. 292 (Vol. 5, p. 145).

[18] Du Moulin, C., *Tractatus commerciorum* ... , Paris (1546) (Goldsmiths'-Kress Library of Economic Literature, item 38.1), pp. 209–11.

[19] Santerna, P., *De assecurationibus et sponsionibus mercatorum*, Venice (1552), Cologne (1599); see Maffei, D., *Il giureconsulto portoghese Pedro de Santarém autore del primo trattato sulle assicurazioni (1488)* (Separata do numero especial do Boletim da Faculdade de Direito de Coimbra, Braga (1983), pp. 703–728).

[20] Soto, D., *De Iustitia et Iure* bk. VI q. 7, Lyons (1569), p. 207 (Goldsmiths'-Kress, item 145.5); cf. B. Straccha, *De Assecurationibus*, Venice (1569) Praefatio and sections 13, 46, 79–80 (Goldsmiths'-Kress, item 145.4); de Roover, R., 'The scholastics, usury and foreign exchange', *Business History Review* **41**, 257–271, at 269 (1967).

[21] *Oeuvres Complètes de Huygens*, The Hague (1893 etc), Vol. 7, pp. 95–8; see also Houtzager, D., *Hollands lijf- en losrenteleningen voor 1672*, Schiedam (1950).

[22] Lessius, *De Iustitia et Iure*, bk IV, Louvain (1605), p. 310, quoted in Coumet, E., 'La théorie du hasard est-elle née par hasard?', *Annales* **25**, pp. 574–598, at 592 (1970).

[23] Bauny, P., *Somme des péchez qui se commettent en tous estats, de leurs conditions et qualitez*, Paris (1653), p. 227, quoted in Coumet, p. 592 n. 1; cf. p. 591.

[24] J. de Lugo, *De Iustitia et Iure* disp. XXXI sect. vii, Lyons (1652), Vol. II, p. 447.

[25] Fermat-Pascal correspondence, in Tannery, P., and Henry, C., (eds.), *Oeuvres de Fermat*, Vol. 2, Paris (1894), pp. 288—314, trans. in D. E. Smith, (ed.), *A Source Book in Mathematics*, New York (1959), Vol. 2, pp. 546—565, at pp. 546—7.
[26] *Oeuvres Complètes de Huygens*, vol. 14, pp. 52—95; partial English trans. in Fredudenthal, H., 'Huygens' foundations of probability', *Historia Mathematica* 7, pp. 113—7 (1980).
[27] *Oeuvres*, 60—1; cf. Hacking, I., *The Emergence of Probability*, Cambridge (1975), ch. 11.
[28] Easton, J. B., 'Jan de Witt', *Dictionary of Scientific Biography*, New York (1976), Vol. XIV, pp. 465—7.
[29] Hacking, first contents page.
[30] Leibniz, *New Essays* bk. 4 ch. 2 in P. P. Weiner (ed.), *Leibniz, Selections*, New York (1951), pp. 82—3; see Schneider, I., 'Leibniz on the probable', in J. W. Dauben (ed.), *Mathematical Perspectives*, New York (1981), pp. 201—219.
[31] *Corpus Iuris Civilis, Digest* 30.96.
[32] *Dig.* 35.1.36.1; cf. 32.64 and 50.5.56.
[33] *Dig.* 50.10.1.1.
[34] *Dig.* 12.4.6.
[35] *Dig.* 23.4.26.
[36] *Dig.* 37.10 and 43.4.3.
[37] *Dig.* 4.2.23.
[38] *Dig.* 41.3.44.4.
[39] *Deuteronomy* 17:6; 19:5; *Numbers* 35:30; *Matthew* 18:16; *Corpus Iuris Civilis, Code* 4.20.8.
[40] Azo, *Lectura super Codicem* bk 4 tit 1, Paris (1577), repr. Turin (1966), p. 254; cf. Ioannes Fasolus *De Summariis Cognitionibus* ed. L. Wahrmund, Aalen (1962), p. 21; Ioannes Andreae, *Novella Commentaria in secundum Decretalium* tit. xix rubric, Venice (1581), repr. Turin (1963), p. 109A.
[41] *Decretals of Gregory IX*, bk. V tit. xxxix cap. 44 (in A. Friedberg, (ed.), *Corpus Iuris Canonici*, Leipzig (1879), Vol. II, col. 908).
[42] *Corpus Iuris Civilis, Digest* 48.18.1.
[43] Johannes de Lignano, *Super Clementina 'Saepe'*, (ed.) L. Wahrmund, Aalen (1962), p. 9.
[44] Thomas Aquinas, *In IV Sententiarum*, dist. 9 q. 1 art. v solutio II.
[45] Thomas Aquinas, *Summa Theologiae*, II—II q. 69 art. 2.
[46] Lottin, O., 'Le tutiorisme du treizième siécle', *Recherches de Théologie Ancienne et Médievále* 5, pp. 292—301 (1933).
[47] T. Deman, 'Probabilisme', *Dictionnaire de Théologie Catholique* Vol. 13, part 1, Paris (1936), cols 417—619; J. de Blic, 'Barthélémy de Medina at les origines du probabilisme', *Ephemerides Theologicae Lovanienses* 7, pp. 46—83, pp. 264—291 (1930).
[48] *Constitutio Criminalis Carolina*, sections 18, 22, 24—30, trans. in J. H. Langbein, *Prosecuting Crime in the Renaissance*, Cambridge, Mass (1974), pp. 261—308, at pp. 272—5.
[49] Menochio, G., *De praesumptionibus, conjecturis, signis et indicijs commentaria*, Venice (1587—90).
[50] Alciati, A., *Tractatus de praesumptionibus*, Lyons (1561); Cologne (1580).

[51] Mascardi, G., *Conclusiones probationum*, or *De probationibus*, Venice (1584—88); Venice (1593); Frankfurt (1593) etc.
[52] Coke, E., *The First Part of the Institutes*, 6b; cf. *English Reports*, Vol. 77, 534; Vol. 145, 73; Shapiro, B., *Probability and Certainty in Seventeenth Century England*, Princeton (1983), pp. 177—8.
[53] Meulenbroek, B. L., *Briefwisseling van Hugo Grotius* Vol. 5, The Hague (1966), p. 194; Haggenmacher, P., *Grotius et la Guerre Juste*, Paris (1983) p. 495.
[54] Grotius, *Law of War and Peace* bk. 2, ch. xxvi (*Rights of War and Peace*, London (1682), p. 430); see Johnson, J. T., *Ideology, Reason and the Limitation of War*, Princeton (1975), p. 220.
[55] Haggenmacher, p. 196.
[56] Grotius, *De Iure Praedae Commentarius*, (trans. G. L. Williams, Oxford (1950), pp. xiii—xv); see Haggenmacher, pp. 200—203.
[57] *De Iure Praedae*, trans. Williams, pp. 80—82.
[58] Chillingworth, W., *The Religion of Protestants a Safe Way to Salvation*, ch. iv, par. 57 (in *The Works of William Chillingworth*, Philadelphia (1940), p. 296); but see also Bellhouse, D. R., 'Probability in the sixteenth and seventeenth centuries: an analysis of Puritan casuistry', *International Statistical Review* **56**, pp. 63—74 (1988).
[59] Chillingworth, ch. vi, par. 5, pp. 430—1.
[60] Hacking, ch. 8.
[61] E.g., Azo, *Lectura super Codicem* (Paris, 1577, repr. Turin, 1966), p. 283.
[62] Hacking, pp. 77—8.
[63] Schneider, pp. 202, 204; Hacking, ch. 10.
[64] Leibniz, *Philosophischen Schriften*, ed. C. I. Gerhardt, Berlin (1875—90), Vol. VII, p. 167; see Hacking, p. 86.
[65] *Philosophischen Schriften*, Vol. III, p. 194.
[66] Biermann, K.-R., and Faak, M., 'G. W. Leibniz, 'De incerti aestimatione', *Forschungen und Fortschritte* **31**, pp. 45—50 (1957); see Schneider, p. 207.
[67] Leibniz, Letter to Wagner of 1698; Schneider, p. 203.
[68] Schneider, p. 210.
[69] See, for example, Eggleston, R., *Evidence, Proof and Probability*, 2nd ed., London (1983).

KIRSTEN BIRKETT AND DAVID OLDROYD

ROBERT HOOKE, PHYSICO-MYTHOLOGY, KNOWLEDGE OF THE WORLD OF THE ANCIENTS AND KNOWLEDGE OF THE ANCIENT WORLD

Perhaps Bacon's greatest influence on the seventeenth-century intellectual upheaval was his insistence on observation as the basis for knowledge of nature. This went hand in hand with his rejection of ancient authority as the source of knowledge. It is easy to glean a picture of the wind of empirical inquiry blowing away dusty tomes of ancient knowledge. This, however, like all simple pictures, is not adequate. It overlooks the influence that Antiquity still held over seventeenth-century thinkers. While the Scientific Revolution was certainly a revolution, the effect of centuries of relying on the past could not vanish overnight. Certainly the new science was empirical and experimental; yet arguments from Antiquity were still used. This is exemplified in the work of Robert Hooke, who was ready to see the advantage of having the Ancients on his side.

We wish to show how Hooke made use of ancient texts in order to develop, and attempt to defend, his ingenious theory of the Earth, designed to explain the presence of the remains of marine organisms in strata now distant from the sea. This use of ancient texts was in accord with a trend of interpretation of mythology in this period. Hooke, however, was unique in that he carried out his interpretation specifically to support his science. To our knowledge, this use of ancient sources as a means of attempting to verify empirically a modern scientific hypothesis was the first significant example of its kind,[1] apart from the use of ancient astronomical observations to attempt to determine quantities such as the rate of precession of the equinoxes. Thus, for the first time, the ideas of the Ancients were mustered not just as sources of scientific theory, but for the empirical information they might furnish to help verify a theory. In the event, Hooke's use of ancient sources did *not* yield evidence sufficient to persuade his critics that his theory was satisfactory; and his use of ancient mythologies for the development of a theory of the Earth turned out to be an unsuccessful method for reasoning about the Earth's geological history. Even so, it provides an insight into the way in which Hooke sought to explain geological phenomena: and the *failure* of his method is instructive. Hooke's testing was done chiefly with the help of astronomical instruments and ancient texts. Ensconced in London, he did not go out into the field to seek to confirm or deny his remarkable hypotheses. Thus he cannot be seen as one who established the science of geology, for which *field work* is an essential component.

Though this paper is not strictly concerned with Hooke's theory of the Earth — we are interested rather in his Classical scholarship, his sources, and the uses he made of them — a few words must be said about the nature of his 'geological' ideas, and how they developed during the course of his career. In the south of England, where Hooke was reared, there are strata containing the petrified remains of organisms that look like marine shells. Many of them, however, are different from those found in modern seas; and the strata where such fossil shells are found may be considerable distances from the sea, or even in some parts of the world (other than Britain) at the tops of mountains.

The question of how these fossils came to be where they are found was much discussed by seventeenth-century natural philosophers, some of whom (such as Robert Plot, Martin Lister and Edward Llwyd) questioned the organic origin of the fossils. Hooke, however, firmly believed that the fossils represented the remains of living organisms, and to explain their presence he hypothesized that since Creation the Earth had often changed; in particular, parts which are now dry land were previously under the sea, and *vice versa*. Hooke held that such catastrophic changes were brought about, or accompanied, by earthquakes.[2]

Hooke's theory was first presented to the Royal Society in a series of lectures given in 1667 and 1668. By 1687 Hooke had developed this into an ingenious theory. The Earth's poles, he suggested, are gradually shifting. This pole wandering could have been a cause of the great earthquakes Hooke was proposing. Also, the changes in the position of the poles with respect to the Equator would have meant that land moved gradually in and out of the sea waters, producing thereby alternating periods of deposition and erosion.[3] Hence the presence of inland shells would be explained. This expanded theory was presented to the Royal Society in 1686 and 1687; and as has been shown by A. J. Turner,[4] Hooke's ideas, as stated in his discourses of 1686/7, were transmitted by Edmond Halley to the members of the Oxford Philosophical Society, led by John Wallis, where they were discussed early in 1687. Apparently the ideas were not well received, for it was felt by the

Oxonians that Hooke's suggestions were incompatible with what was known of the ancient history of the world. The Oxford response was transmitted to Hooke in the form of a letter from Wallis to Halley, and this led to an acrimonious exchange between Hooke and Wallis which has been analyzed elsewhere[5] and need not concern us here.[6] What is interesting to us at present is that, in responding to Wallis, Hooke began a close search of ancient texts in order to try to muster support for his theory. Thus, in his closing years, Hooke turned to the evidence proffered by ancient writers, such as Plato, Ovid, and Pliny, in order to shore up his theory, and he gave a number of lectures on the topic to the Royal Society. It is this work that we wish to scrutinise, to see how Hooke sought to enlist ancient writers in his cause. We shall be interested in how he used the sources, the quality of his scholarship, and the ultimate success or failure of his enterprise. The texts to be analyzed are all found in his *Posthumous Works*, published by Richard Waller in 1705, on the basis of manuscripts collected together after Hooke's death.[7] Waller did not do a thorough editorial job, and the lectures are not presented in their correct chronological order. It is only recently that the chronology of Hooke's geological writings has been established by Rhoda Rappaport.[8]

Apart from Dr. Rappaport's work, Hooke's interest in ancient mythologies has been largely ignored or treated superficially. Indeed, early commentators on his work regarded it as almost aberrant, and indicative of a decline of his intellectual power.[9] But it is not difficult to see why and how Hooke turned to ancient sources as he did. They were, in fact, a major pre-occupation through much of his career, as can be seen from the analysis of his personal library in Appendix I. About one-third of his 'Discourse of Earthquakes' concerns the examination of ancient texts. So what sources did Hooke use, and just how did he use them?

'A DISCOURSE OF EARTHQUAKES'

The lectures that Hooke gave in 1667/8 set the agenda for the ensuing debate. While the ancient texts were not employed extensively until much later, even in the 1660s Hooke showed his desire to back up his empirical discussion with the words of the Ancients. These sources were treated briefly — almost in passing — but already it was possible to see that Hooke recognized a need to have historical support for his geo-historical theory. The ancient writers mentioned were Pliny, Strabo, Seneca, Virgil, Plato, Ovid and Scripture. For our analysis of Hooke's use of these source, see Appendix II.

The matter was taken up again in 1686. Unfortunately, Hooke acknowledged, there were few histories available about the geological past of the Earth, since the 'great transactions of the Alteration, Formation, or Dispositions of the Superficial Parts of the Earth'[10] occurred before the invention of writing. Consequently, in his lectures later in 1686 and 1687, Hooke sought to provide direct empirical evidence, in the form of records of changes in the direction of the meridian in London, and elsewhere.[11]

By December 1687, however, it seems clear that this alternative source of information was not convincing Hooke's audience. 'One of the most considerable Objections I have yet heard,' he wrote, 'is that History has not furnish'd us with Relations of any such considerable changes as I suppos'd to have happen'd in former Ages of the world.'[12] This, as we know from Turner's work,[13] was the chief objection raised by the members of the Oxford Philosophical Society.

So here began in earnest Hooke's search into history and mythography — a search which was to continue virtually until the end of his life. Four sources particularly featured and provided the basis for his argument: Ovid's *Metamorphoses*; Plato's *Timaeus*; Hanno the Carthaginian's *Periplus*; and the Bible. We shall now look at these in turn, to see what use he made of these texts, and to consider whether they might have carried the weight that Hooke sought to place upon them.

PLATO, HANNO AND ATLANTIS

Hooke had Plato in mind from the beginning of the 'Discourse'. In 1667, he mentioned Plato's *Timaeus*, and the story of Atlantis in particular, as supporting the idea that land may be sunk beneath the sea by an earthquake. However, there was no further discussion in 1667, only the name of the account being mentioned. It was on 7th December 1687 that Hooke expanded his thoughts about Atlantis. Though some people, Hooke said, might think the account fictional — no more that a setting for the *Republic* — there was so much probability in it that at least the story showed that Plato had some notion of preceding states of the world; and good reason for his belief. That is, Plato 'did suppose and believe that there had been in many preceding Ages of the World, very great changes of the superficial Parts of the Earth by Floods, Deluges, Earthquakes, etc.'[14]

As further evidence, Hooke went on to relate the history of the *Periplus* (that is, 'Voyage') of Hanno the Carthaginian, who led an expedition beyond the Columns of Hercules (Straits of Gibraltar) to build 'Lybyphenician'[15] settlements on the north-west coast of Africa. The text[16] described this voyage and the various lands supposedly encountered. The salient part of the story, so far as Hooke was concerned, was that Hanno's fleet seemed to have passed through the area where Atlantis was presumed to have sunk — that is, to the west of the north-African coast — and found various lakes and islands and an active volcano. What was more, to the best of Hooke's knowledge, the same area contained no such features in his day; so, he argued, land must have sunk if Hanno was to be believed. All in all, Hooke regarded this as evidence in favour of his notion 'that there have been in those parts prodigious alterations somewhat like those I have supposed in my Hypothesis'.[17] What appeared in Hanno's account to have been a former volcano might, Hooke plausibly suggested, be represented by the Pike of Teneriffe in the Canary Islands.[18]

Plato was discussed again on 15th February, 1688. It appears that although Hooke had, by reference to Plato, answered the objection that histories gave no accounts of catastrophic changes, the evidence from Plato's text was not regarded as sufficient, the story of Atlantis being apparently held in some doubt by his auditors.[19] In answer, Hooke listed other ancient authors who apparently corroborated Plato. Hooke called on the authority of Strabo and Eratosthenes to uphold the view that Plato's history of Atlantis was not fiction. (The passage is a little unclear.[20]) Also, Pliny's *Natural History* was invoked in support for Plato;[21] and Pliny's own description of earthquakes was cited in defence of Hooke's hypothesis.[22]

Hooke went on from Plato to describe once again how the *Periplus* offered evidence: but Hanno too, it appears, had come under attack from Hooke's auditors.[23] Consequently, Hooke listed other ancient authors who lent credibility to Hanno's story. Strabo reported that the 'Sidonians' were good philosophers, astronomers and arithmeticians;[24] another author (taken to be Aristotle) said the Carthaginians had discovered a deserted island beyond the Pillars of Hercules.[25] Pomponius Mela,[26] Hooke said, mentioned that the Atlantic was inhabited by a wild people, and a kind of Satyrs. Diodorus Siculus also referred to an island in that part of ocean.[27] All of these, Hooke claimed, supported his reading of Hanno.

How acceptable was it for Hooke to use the story of Atlantis as evidence for his scientific theory? It seems that some credence was given to the Atlantis story in the seventeenth century, though this was rapidly fading. Interest in the myth had been heightened with the Renaissance exploration of the world, and various theories about Atlantis surfaced during the sixteenth century. The polymathic Jesuit scholar Athanasius Kircher in 1665 literally put Atlantis on the map, and regarded the Azores and the Canaries as its residual remains.[28] It seems that it was by no means ridiculous for Hooke to quote Plato for support. However, it is clear that this evidence was not regarded as conclusive. This comes out in the constant defence that Hooke provided for Plato — it would seem that Hooke's hearers raised some fairly strenuous objections to this form of argument. Yet by looking for textual evidence corroborating Hanno and Plato, Hooke was behaving like a modern critical historian, even though he ignored the possibility that his 'authorities' were not necessarily independent: one could have acquired his information from another, and very likely did so. And as will now be shown, Hooke was fanciful, almost gullible, rather than critical, in his construals of the mythological writings of Ovid.

OVID

Hooke's theory of ancient mythology was that the myths which had been described by ancient authors in terms of seemingly impossible human (or divine) actions were in fact records of catatrophic events. The Ancients, he suggested, would have preserved some histories or traditions of such catastrophes which would have been remembered in myth to conceal such knowledge from the vulgar. Only the initiated would have the key to the stories; these stories would, then, entertain the vulgar while serving to instruct the learned. Hesiod's *Theogonia*, Hooke claimed, was of such a sort: it seemed to contain the earliest histories that the Greeks obtained from the Egyptians or Phœnicians.[29]

This is how Hooke construed Ovid. Ovid, he claimed, studied to inform himself of the 'Changes and Catastrophies that had happened to the Earth from the Creation unto his own time'.[30] He gathered this information together in his *Metamorphoses* under the guise of the doings of gods and giants. The first four verses of the *Metamorphoses* seemed to Hooke to state this clearly:

> I sing of Beings in new shapes array'd,
> Assist ye Gods (for you the Changes made,)
> That from the Worlds Beginning to these Times
> I may comprize their Series in my Rimes.[31]

'Which is as much to say', Hooke claimed (in a *very* liberal interpretation): 'My [Ovid's] design in this Book is to speak concerning the various alterations and transformations which the Bodies or superficial Parts of the Earth have, by the Divine Powers, undergone'.[32]

Hooke, therefore, perceived his task to be to 'disinterpret' Ovid in order to glean the 'true' geological information behind the myth. This procedure he undertook over several lectures throughout the period of debate. Rearranging the sequence in which Hooke presented them, it is possible to see (in the order in which they occur in the *Metamorphoses*) Hooke's interpretation of several of Ovid's fables.

The first part of the *Metamorphoses* was quite straightforward — it described the formation of the Earth, then the lakes, seas, rivers, hills, and so on. According to Ovid's mythology, the history of the world was divided into four ages (which Hooke claimed were postulated by Varro[33]): Golden, Silver, Bronze (which Hooke did not consider) and Iron. It was during the Iron Age that the fables in which Hooke was interested took place, and he proceeded to give his 'geological' interpretation of a number of the particular stories in the *Metamorphoses*. (For a complete list, see Appendix II.) His general assumption was that most of the curious events recounted by Ovid were to be understood as referring obliquely to specific geological phenomena, or events in the Earth's history (usually of catastrophic nature) actually seen by humans.

Hooke was undoubtedly taking an unusual tack in this interpretation of Ovid. It is useful to look briefly at the history of interpretation of the *Metamorphoses* to see how this work had previously been read.[34] Ovid was available from Antiquity right through the Middle Ages and into Hooke's time. In the Middle Ages, Ovid's work was seen as a moralistic text and as theological allegory. During the Renaissance, while the moralizing and allegorical interpretation continued, the approach was changing. A more naturalistic tradition of interpretation arose, the gods being seen as parallels for physical things. This was the beginning of a trend towards a more historical interpretation, away from the moralizing. Arthur Golding, who produced the first English translation of Ovid in 1567,[35] while his commentary was still largely moralistic and endeavoured to reconcile the pagan work with scripture, gave some naturalistic explanations of the text. George Sandys, who published a widely-used translation of Ovid in 1632, offered interpretations that were even more naturalistic.[36] Other commentators offered elaborate astronomical/astrological readings.[37]

There is, however, no evidence that anyone prior to Hooke interpreted Ovid as geological history. Certainly Hooke was careful about the explicit weight he placed on his interpretation (despite the fact that this interpretation was a large part of his argument). He admitted that some people might take different interpretations of Ovid, moral or even Biblical. He thought his the most plausible — while acknowledging that poets couched stories in terms that gave a physical and moral, as well as a historical, meaning. So even if a moral meaning could be gleaned from Ovid, the text was still for Hooke primarily about the Earth's history.[38] Hooke was convinced that his interpretation was right. Yet he seemed aware that others did not share his views, and at times he seemed to be more apologetic than insistent.[39]

SCRIPTURE

As early as 1667, Hooke used Scripture to support his theories. Passages in the Bible were cited to back up various points during the course of his lectures. The command for the waters to go to their place would, he claimed, have been actuated by tremendous earthquakes, so great alterations certainly took place;[40] it also seemed likely to Hooke that Noah's Flood greatly changed the Earth, as an earthquake must have been the mechanism by which the sea was made to cover the land;[41] the cities of Sodom, Gomorrha, Zeboim and Adma were also probably destroyed by earthquakes.[42] It is interesting that these theories of the mechanism of Biblical events were quite as imaginative as Hooke's interpretation of Ovid.

Hooke's views on the Creation and the Flood were explained most fully, and just as imaginatively, on 29th February 1688 when he gave a model of the Earth as he saw it described in Genesis 1:2, and the changes that must have taken place during the six days of the Creation. The Earth apparently remained in this state until the time of the Flood (sixteen to seventeen hundred years later). Hooke went on to relate the way in which the flood occurred, giving an account which was conjectural, but which fitted the Biblical narrative and provided a plausible explanation for the phenomena of sea shells on dry land. (See Appendix III for Hooke's account.) The theory that Hooke was here advancing, on the basis of Biblical authority, was different from the scheme proposed in his earlier work,[43] which was more obviously naturalistic, and working in accordance with general mechanical principles. Thus he was endeavouring to propound a hypothesis that reconciled Scripture and pagan history, as well as his own earlier 'mechanical' scheme. It was not an easy task.

Later Hooke used various Bible passages as evidence that the Earth had changed considerably since Creation. The Earth, he claimed, was totally changed during the Flood (Gen. 6:13; 8:21; 2 Peter 3:5,6).[44] Hooke also found in Scripture evidence that the world was always changing and growing old. He quoted Psalm 102;[45] Isaiah 51:6; and Hebrews 10—12 was mentioned.

Hooke was writing in an era which was beginning to look at the Bible through scientific glasses. Protestantism had sent scholars back to finding the real meaning of the Bible — there was far more interest in the original texts. The discovery of older manuscripts of the Bible led to a more accurate version being available and this was becoming ever more widely read with the help of printing. At the same time, new historical information was accumulating which had to be fitted into the Bible narrative. There were problems of Antiquity, as demonstrated by Egypt; also the practical problems of discovery of new flora and fauna — how did they all fit in the Ark? Science was called upon to solve such problems.

However science, originally brought to the defence of Scripture in the light of new information, proved to be a difficult ally. Many people were wrestling with the problem of bringing God's Creation into line with God's Revelation:[46] for example Thomas Burnet,[47] Erasmus Warren,[48] William Whiston,[49] Athanasius Kircher[50] and Sir Matthew Hale.[51] Some denied that fossils were of organic origin at all — a considerable difficulty being the time required for their emplacement. Those who accepted their organic origin — such as John Woodward — looked to the Flood to solve the problem of distribution.[52] Such 'reconciliations' of science and religion, in which science and theology become mutually entwined so as to reinforce one another, constituted the '*physico-theology*' of the latter part of the seventeenth century. The relationship was, nevertheless, an uneasy one, and the ideas of both Burnet and Woodward came in for censure.

Hooke, however, was not using science to defend the Bible; rather, he was using the Bible to defend his science. Yet while conscripting the Bible's help, Hooke managed to provide a reasonably coherent account of the events of the Creation and the Flood. Essentially it was *this* — his theory of the mechanisms whereby the Earth underwent change — which supported him, not the Scripture itself. Without his special interpretation, the Bible would be no help. Yet while he was alone in the particular 'geological' theory that he developed, his 'scientific' treatment of Scripture does not seem at all out of place in this period.

ADDITIONAL SOURCES

Although, as we have said, Hooke's chief ancient sources were Plato, Hanno, Ovid and the Bible, he did mention other authors at times. Herodotus was cited for evidence that the Ancients were aware of the phenomena of shell-like bodies on mountain tops.[53] Aristotle apparently confirmed this notion that Egypt had once been sea,[54] and he was also cited for saying that the boundaries of the seas could change.[55] Ovid was also called upon for his record that Pythagoras noticed the shell phenomenon.[56] Hooke took such observations of the Ancients as a possible source of the many traditional flood stories, and he considered a few of these, but we have insufficient space for further details here.

HOOKE'S USE OF HISTORY AND HIS INTERPRETATION OF MYTH: 'PHYSICO-MYTHOLOGY'

We have examined the way in which Hooke made use of ancient writings in order to find support for his theory of the Earth. What Hooke was engaged in was an interesting form of classical scholarship. His approach may be described as a kind of euhemerism — interpreting ancient mythical writings in terms of natural, non-divine events. However, as he considered these events to be geological in character and not to do with people, rather than use the term 'euhemerism' we have coined the term '*physico-mythology*'. The term is, of course, suggested by the idea of 'physico-theology' — for Hooke was treating mythology in terms of the physical world, making his studies of myth and physical science mutually reinforce one another. We have already given some indication of the history of this euhemeristic approach to mythology, in the particular areas of Plato and Ovid. This overlaps with the area of Biblical scholarship too, for (as will be seen) a great deal of the traditional interpretation of mythology was concerned with the effort to reconcile pagan myths with the Bible.

Euhemerism began with Euhemeris of Messene (ca. 300 B.C.) in his work *Sacred Writing*. Here Euhemeris propounded the theory that the gods of mythology were actually humans who had come to be spoken of as deities. This idea was taken up by other Classical writers, such as Diodorus Siculus and Palaephatus,[57] who both wrote of the humans who inspired the gods of myth. The more general idea of there being a message in mythology different from what appeared on the surface was not an uncommon one in the Classical period, as seen for example in Strabo.[58]

In the Christian era, euhemerism was taken up by apologists wishing to undermine pagan religion:[59] this gradually developed into the numerous chronologies and attempts to explain myth as corrupted Scripture (Sir Walter Raleigh's history of the world is one example).[60] Meanwhile, other strands had developed from euhemerism. We have already seen the moralistic interpretation traditionally used for Ovid, developing into a more historical interpretation. Similar notions of myth were held by other seventeenth-century writers. Francis Bacon wrote a work on myth, *Of the Wisdom of the Ancients* (1605). Agreeing that there must be more to myth than meets the eye,[61] Bacon went on to give his own interpretation of numerous myths. These interpretations were a far cry from Hooke: Bacon's interpretations were largely moral/political. For example, the story of Typhon, which Hooke claimed was about earthquakes, Bacon described as a political revolution.[62] Yet Bacon was careful to qualify his enthusiasm, realising that it is easy to read one's own philosophy into fable.[63]

Other seventeenth-century writers tried their hand at a theory of interpretation of myth, some more naturalistic, some more moralistic. Henry Reynolds, for example, in his *Mythomystes* (1632),[64] was convinced, as was Hooke, that the Ancients took care to hide their wisdom in myth. Reynolds, however, was another who came nowhere near Hooke in his actual interpretation; while he did speak of natural events, they were very general. It is hard to tell whether there was any 'official' view about myth in the Royal Society. Sprat did mention the matter in his *History of the Royal Society*, though only in passing. He was quite dismissive:

The *Wit* of the *Fables* and *Religions* of the *Ancient World* is well nigh consum'd: ... and it is now high time to dismiss them; especially seeing they have this peculiar *Imperfection*, that they were only *Fictions* at first.[65]

In general, the seventeenth-century view seems to have been that myths were degenerate forms of Biblical truth.

It remains to show how Hooke fitted into this tradition.[66] It seems that to search for 'hidden meaning' in ancient texts was not out of line — this followed in a long tradition of interpretation, even if the view of the Royal Society may have been more sceptical. However, we would emphasize that Hooke was not so much following an established tradition, as using it as a vehicle for creating evidence for his theory. For Hooke, to look into historical records for evidence showed quite a canny sense of what was needed: after all, he was convinced (on the basis of the evidence of the shells) that major geological changes must have happened in the ancient past, so he was looking in the right place for his evidence. However, there is no suggestion that anyone else in the seventeenth century found descriptions of earthquakes in ancient writings, as Hooke did. He was not so much trying to discover the texts' meanings, as trying to find texts which could possibly support the meaning he placed in them. (Notice that at the beginning of 1686 Hooke did not expect the Ancients to be of much use: it was only after his astronomical experiments failed that he turned to their writings.[67]) This would explain the sometimes rather cavalier scholarship — the occasional misquotes and imaginative readings. Certainly Hooke's interpretation of Ovid was a triumph of imagination. (See Appendix II.)

Ultimately, Hooke's attempt was not successful: but his work makes an interesting comment on the place for ancient texts in seventeenth-century science. They were not accepted as authority in the end — but then, Hooke was not really using them as authority for his own theory, as Copernicus had referred to the ideas of the Pythagorean, Philolaus, as an ancient precedent for the idea of a moving Earth.[68] He was using them as any other record, as evidence that certain events had occurred. It was his interpretation of the texts that was rejected, not the 'wisdom of the Ancients' that was being denigrated, though some of Hooke's listeners were obviously sceptical of the veracity of the texts themselves. (Perhaps Sprat's quote throws some light on this.) What is important is that Hooke did go to the ancient writers for evidence; and while in the end he was defeated, he was certainly never laughed out of court.

CONCLUSION

The way in which knowledge was established in the early Royal Society has been the subject of recent comment. Steven Shapin[69] has described the pattern of turning opinion into knowledge as including a (semi-) public exhibition. Experimental phenomena, already tried in private (the real research), were shown or demonstrated in front of the Royal Society, and discourse about the experiment could ensue. Not until this process of showing in a public space (demonstration) had been carried through could an experimental phenomenon become accepted as 'knowledge'. Claims made in the semi-public arena of the meeting chamber of the Royal Society were evaluated by the Fellows. When agreement was reached, the 'knowledge-claims' were then published, with the Society's imprimatur. And thus they became 'knowledge' rather than 'knowledge claims'.

Another important issue, however, was that of testimony.[70] Experience was reported to the Royal Society, and the validity of the knowledge claim rested on the authority of the witness. In a sense, Hooke was searching for authoritative 'witnesses', or at least recorders, of events, which would provide evidence for his theory. But his geological ideas could not be demonstrated in the try/show/discourse/publish pattern: the experiments he performed or intended, observing stars for secular changes in the meridian and so on, could hardly be demonstrated at a meeting of the Royal Society. He was relying on the testimony of ancient sources, which were acceptable allies, though not sufficiently so for his theory to be believed. The 'Moderns' were definitely beginning to take over from the 'Ancients': yet, as the example of Robert Hooke shows, ancient knowledge still had a place in seventeenth-century science.

ACKNOWLEDGEMENTS

We are indebted to Rhoda Rappaport for her detailed and helpful referee's report; to Stephen Gaukroger for his editorial assistance; and to Jamie Kassler for a copy of the sale catalogue of Hooke's library.

APPENDIX I

An Approximate Subject Analysis of Hooke's Personal Library[71]

Topics	Numbers of Books
Mathematics/Geometry/Algebra/Arithmetic	299
Medicine/Health/Anatomy/Surgery/Midwifery/Pharmacopœias	276
Geography/Travel/Maps/Topography/Descriptions of Cities	265
Astronomy/Astrology/Cosmogony	239
Theology/Bibles/Liturgy/Catechisms/Missionaries/Different Churches	200
Classics[72]	192
Miscellaneous[73]	154
History/Historiography/Mythography	121
Etymology/Language/Grammar/Philology	110
Poetry/Plays/Epigrams/Colloquies	102
Memoirs/Biography/Autobiography	80
Botany/Gardening/Silviculture/Agriculture/Floristry	76
Chemistry/Alchemy	63
Morals/Ethics/Maxims	56
Recreation/Courtiership/Etiquette/Manners/Decorum/Conversation/Fortune/Love	55
Philosophy/Natural Philosophy/Science (general)	54
Theories of the Earth/Minerals/Gems/Stones/Metals/Mining/Volcanoes/Earthquakes	51
Law/Politics	48
Physics (general)	47
Method/Logic/Reasoning/Mind/Memory/Thinking	46
Letters/Correspondence	44
Light/Optics/Microscopy/Telescopes	42
Fortifications/Military/Tactics/Warfare	40
Painting/Drawing/Art/Artists/Miniatures/Sculpture/Engraving/Perspective/Glasswork	40
Machines/Mechanics/Dynamics/Gunnery	38
Dictionaries/Lexicons/Thesauri/Encyclopædias	36
Mystical Writings/Occult/Arcane/Secrets/Witchcraft	34
Architecture/Building	25
Animals/Birds/Insects	24
Natural History	23
Chronology/Calendars	22
Mensuration/Weights and Measures	22
Fiction	23
Medicinal waters/Fountains	18

Trade/Business/Economics/Wealth/Coinage	18
Libraries/Library catalogues/Bibliography	17
Life/Death/Longevity	16
Eastern philosophy/Orientalism	15
Music/Accoustics/Songs	15
Curiosities/Rarities	15
Learned journals	15
Orations/Oratory/Rhetoric	14
Aristotle/Aristotelianism	13
Navigation	13
Origin and nature of man	13
Horology	12
Hydrostatics/Hydraulics	12
Antiquities	11
Meteorology/Winds	11
Education	10
Satire	10
Inventions	9
Surveying	9
The Creation/Pre-Adamites/Universal flood	9
Cryptography	7
Hieroglyphics/Egypt	7
Magnetism	7
Natural Magic	7
Crystals/Salts	6
Fermentation/Wine	6
Husbandry	6
Naval history/Shipbuilding	6
Proverbs	6
Shorthand	6
Art of writing	5
Apiculture	4
Cooking/Dietetics	4
Dialling/Sundials	4
Fireworks/Pyrotechnics	4
Amber	3
Humour (venal)	3
Epicureanism	3
Tobacco/Beverages	3
Dyestuffs	2
Generation	2
Heraldry	2
Hygrometry/Barometry/Thermometry	2
Ice/Snow/Cold	2
Koran	2
Metaphysics	2
Mortality bills/Political arithmetic	2

160   KIRSTEN BIRKETT AND DAVID OLDROYD

Riding/Dancing	2
Ghosts	1
Girls' schools	1
Differences in intelligence	1
Interest tables	1
Origin of writing	1
Palmistry	1
Physiognomy	1
Polygamy	1
Printing/Typography	1
Prostitution	1
Silk/Sericulture	1
Teaching birds to speak	1
Tides	1
Time	1
Torturing practices	1
Total	3296

APPENDIX II

Ancient Sources Discussed in the Lecture Series Beginning 27th June 1667

Reference as given by Hooke	What Hooke did with the source	Notes
Pliny		
Natural History Book 35, Chapter 13	Quoted in Latin: no English translation. Examples of earth being converted into stone due to the effects of sea water and air. (*P.W.*, 296)	In the Loeb edition of Pliny's *Natural History*, this is in Chapter 47, 385.
Natural History Book 31, Chapter 10	Quoted in Latin: no English translation. Describes how nitre [natron] becomes stone in the regions around Memphis. (*P.W.*, 297)	Chapter 46, Loeb edition, 447.
Natural History Book 2, Chapter 87	Claims that land may be raised by earthquakes. (*P.W.*, 299)	Reference same as Loeb edition, 331.
Natural History Chapter 2, Chapters 86 and 87	Quoted in Latin: no English translation. An instance of part of the sea bed being raised by an	Chapters 88 and 89, Loeb edition, 331. These islands

ROBERT HOOKE AND PHYSICO-MYTHOLOGY 161

	earthquake to form islands, including Anaphe, Thera and Therasia. (*P.W.*, 300—301)	are near Crete.
Natural History Book 2, Chapter 88	Quoted in Latin: English translation following. The mountain Epopon was levelled and a town swallowed, leaving a lake. (*P.W.*, 307)	Chapter 89, Loeb edition, 333.
Natural History Book 2, Chapters 90, 91 and 92	No quotation: Hooke says Pliny gives instances of mountains and hills sinking into plains and lakes due to earthquakes. (*P.W.*, 307)	Reference same as Loeb edition, 335—66.
Natural History Book 2, Chapter 48	Quoted in Latin: English translation. The greatest earthquake happened during the reign of Tiberius Caesar: twelve Asiatic cities were thrown down in one night. (*P.W.*, 310)	Chapter 86, Loeb edition, 331.
Natural History Book 2, Chapter 83	Quoted in Latin: English translation. Two hills collided with a great crash, flame and smoke; many animals and houses perished. Also in the last year of Nero another such diaster occurred. (*P.W.*, 310)	Chapter 85, Loeb edition, 330—31.
No reference	The Alps and Appenines are often troubled by earthquakes. (*P.W.*, 311)	Book 2, Chapter 82, 325.
Strabo Tenth Book	No quotation. Hooke says Strabo mentions the island (Anaphe) whose formation was described by Pliny. (*P.W.*, 301)	Book 10, Chapter 5.1. Strabo merely mentions that Anaphe exists; he does not speak of how it was formed.
Seneca *Natural History* Book 6, Chapter 21	Quoted in Latin. Confirms Pliny's account of Thera and Therasia being raised. (*P.W.*, 301)	Seneca actally claims that these islands were formed by air, or the wind.

Natural History Book 6, Preface	Lengthy quotation in Latin: English translation. A rather philosophical passage, talking of the impermanence of all things; the Earth is movable, cities may sink. (Tyre was destroyed; Asia lost twelve cities; Achaia, Macedonia and Campania have felt the same power.) (*P. W.*, p. 311)	Book 6, Chapter 1, 11—15. The twelve cities were destroyed A.D. 1. Campania was the district around Capua, including Mt Vesuvius: it is not surprising it suffered some devastations.
No reference	Quoted in Latin; a view that Seneca attributes to Fabianus talking of the power of earthquakes to raise the bottom of the sea. (*P. W.*, 314)	Book 3, Chapter 27.[74]
Virgil *Æneid*, Book 3	Quoted in Latin: English translation by Ogilby. A description of the shores of Sicily where the giant Ætna stands. Hooke claimed this is a description of an earthquake. (*P. W.*, 323)	Ogilby (1600—1676) was a writer whose works included translations of Virgil, Homer and Æsop.

Hooke's Interpretations of the Legends in Ovid's Metamorphoses, *1688 and 1693*

Legend	Hooke's Interpretation
The first thing that took place in the Iron Age was the rebellion of the giants. Jupiter had enclosed Saturn in the prison of Tartarus with the other giants. They conspired to break out, and force their way up into heaven, where Jupiter prevailed. They broke forth and made mountains. Jupiter rent the heavens with lightning and buried them at last with mountains heaped on them.[75]	Vapours under the Earth made eruptions and carried the Earth up with them, making mountains. The vapour escaped into the air, producing lightning; eventually this was all over and the mountains remained. *P. W.*, 381.
Lycaon. The blood of the giants produced a new generation; Jupiter grew	Some subterraneous vapours remained, and thunder and lightning ensued. A

angry and called a council of the gods. He had discovered that Lycaon had planned to destroy all living things including Jupiter himself. Jupiter descended and destroyed with fire. Lycaon fled, howling, having changed into a wolf, but still he remained with the human vestiges of grey hair, stern look and fiery eyes. *M*, 7.

Yet some conspirators remained, and men were lost in wickedness. To destroy them all, a universal catastrophe was necessary, so Jupiter decided to flood the Earth. *M*, 8—9.

All were killed by the flood, except for Deucalian and Pyrrha, who were to be the restorers of mankind. They did so by dropping stones behind them. Deucalian's turned into men, and Pyrrha's turned to women. The other creatures came from the Earth, particularly marshes, impregnated by the Sun. Some new creatures also appeared — terrible and destructive monsters, including one particularly venomous serpent called Python, which was eventually killed by the darts of Apollo. *M*, 10.

The story of Daphne (not described in detail). *M*, 14—18.

The story of Io: Juno, Jupiter's wife, found Io in her bed, and out of jealousy set Argus to watch Io and force her to leave. However, Mercury cut off Argus' head, and Io returned to the bed. *M*, 18—21.

The story of Syrinx (not described in detail). *M*, 21—23.

great collection of subterraneous spirits gathered. Lycaon is equivalent to dissolution[76] — that is, these pent-up vapours threatened a general catastrophe that was likely to destroy the Earth. Lightning followed; Lycaon's flight was equivalent to the vapours flying and making noise underground, while the white tops and fiery openings of the mountains remained as a reminder. *P.W.*, 382—83.

There was a flood over all the Earth. *P.W.*, 384—86.

The Sun's heat produced a great growth of vegetation from the moist Earth; but the very boggy places tended to bring forth pestilential and noxious vapours. These were eventually burned off or discharged by the Sun's rays and lightning. *P.W.*, 386—87.

The production of woods and trees by the power of the Sun. *P.W.*, 387.

Juno was the air; Io was the dew, raised into the air as vapour by Jupiter, the Sun. Argus was the stars set to watch by night and cause the dew to fall; Mercury, the light of the morning, cut off the stars' light, and with the Sun the dew was raised back into the air as vapour. *P.W.*, 387.

The generation of water and river plants. *P.W.*, 387.

Juno placed the head of Argus in the feathers of her bird. *M*, 22.	Juno was the air, her bird the clouds: this was the production of the rainbow. *P. W.*, 388.
The story of Phæton. He was a son or production of the Sun and Clymene, the Earth. *M*, 23—24.	Phæton represents fire in the bowels of the Earth. The description of the thundering horses that drew the chariot of the Sun represented lightning and thunder in the air before the production of this fire. *P. W.*, 390 and 392.
Phæton had a great desire to know his father, so Clymene directed him to the palace of the Sun. There Phæton went, and was much pleased with the glorious work, particularly the workmanship of Vulcan on the gates. *M*, 25.	The vapours of the Earth were expelled into the air towads the Sun. The state of the world, however, remained much the same: the Sun still kept its course, the seasons remained, the axis of the world did not change. *P. W.*, 392.
Phœbus described to Phæton the way to take his chariot through the heavens. Phæton, however, did not keep to the path. *M*, 26—30.	Phæton here signified a comet, or fiery meteor, which came close to the Earth and eventually broke into pieces. The description seemed clear to Hooke as signifying great Earthquakes and eruptions in the Earth, to the extent that parts of the sea bottom were raised and made dry. *P. W.*, 393.
Perseus was the son of Jove, begotten in a shower of gold or fire from heaven. He carried with him the Gorgon's head, haired with vipers, which turned all to stone, and which had blood which created snakes. *M*, 93.	Perseus is hot inflamed air, or lightning exhalations from the Earth, set on fire. He was begotten by Jove, æthereal or elementary fire, in a flash of lightning. The Gorgon's hair — vipers — signified lightning. The blood becoming snakes referred to the snake stones, or thunderbolt stones. The gaze of the Gorgon's head referred to the petrifying quality of eruptions. *P. W.*, 397.
Perseus travelled to the country of Atlas (Africa). Atlas knew of a prophecy which said that an offspring of celestial fire would destroy his country; so he had enclosed it with cliffs and a dragon. Yet, Perseus conquered the country by revealing Medusa's head, and Atlas turned into a mountain. *M*, 93—94.	Africa, enclosed by cliffs and the sea, had never been troubled by earthquakes: but now these arrived, and a subterranean eruption destroyed the land, petrifying things to rock and raising a mountain. *P. W.*, 398—400.
The two Gorgons had inhabited certain	The Gorgons signified mountains, with a

islands near Atlas, and had one eye between them. Perseus took the eye, finding Medusa asleep, and cut off Medusa's head. Pegasus and his brother sprang from Medusa's blood. *M*, 97—98.	volcano being the eye; the volcano was not burning at one stage, when subterraneous vapours blew the top of the mountain off. Pegasus and the brother signified new eruptions breaking out. *P.W.*, 400—401.
Perseus then flew to the country of the Ethiopians, where he found Andromeda chained to a rock and expecting to be devoured by a sea monster. Perseus saw her hair waving in the wind, and rescued her. *M*, 95.	Andromeda was part of Africa, a rocky country, by turns covered by water at high tide and sandy at low tide. An eruption raised the frontier of the land, thus repelling the sea and freeing the land from inundation. *P.W.*, 401.
The fable of Proserpine. The giant Typhœus, trapped underground, was shaking the Earth. Pluto, afraid that his residence would be broken, came up out of the Earth with his black horses drawing his chariot. He fell in love with Proserpine, who was innocently gathering flowers, and, seizing her, carried her into the Earth. Ceres, Proserpine's mother, then sought her daughter all over the world. *M*, 109—16.	Proserpine was a town near the lake Pargusa. A great eruption occurred, throwing smoke into the air, and the town was swallowed entirely. Earthquakes continued for some time over a wide region, which Hooke took to be Sicily. *P.W.*, 402—403.

APPENDIX III

Hooke's Explanation of Creation and Noah's Flood

Hooke's model of the earth as it was in Genesis 1:2: the Earth in the centre was a void, with darkness above. Two 'separations' were caused by God: one in the middle of the light, another in the middle of the darkness, so the firmaments of Heaven and Earth were established.

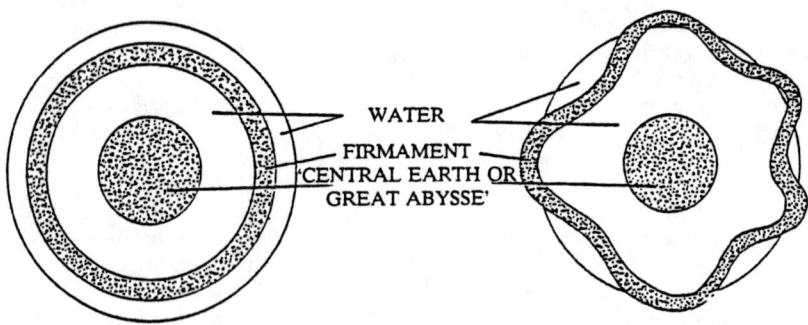

The Earth was now a spherical shell, encompassing the central abyss and a shell of water, with more water forming an outer layer (the oceans). (This model appears to be that held by Ovid, Plato, the Egyptians and the Chaldeans.)

The firmament was then forced outwards in some places, and in other pressed inwards, when (Verse 9) God commanded the waters to be gathered in one place, and dry land to appear (for if one part was pushed in, another part must 'pop out' in balance).

The earth apparently remained unchanged until the Flood. Then God destroyed the Earth: 'the fountains of the great deep were drawn up, and the windows of Heaven were opened'.[77] This, Hooke said, referred to the two-fold firmament — the firmament of the heavens was opened, so it rained for forty days and nights; and somehow water also came from below the firmament of the Earth. The raising of the fountains of the deep Hooke took to mean to mean the raising again of those parts that were during creation sunk to receive the sea — this would have entailed a sinking of the dry land to compensate. The sea would then flow over all the Earth, the rising and sinking surfaces meeting after forty days. Probably this process continued until the Earth recovered its spherical figure, and then continued further until the sinking parts went below level as far as they have been above, and vice versa. Consequently, that which had been the bottom of the sea now became dry land and that which was dry land was now the bottom of the sea. This explains the presence of sea-shells on dry land.

NOTES

[1] Steno looked to Scripture to see that it did not contradict his theory of the Earth: but he was dismissive of other ancient histories as being 'full of fables'. See: N. Steno, *Prodromus to a Dissertation Concerning Solids Naturally Contained Within Solids . . .*, London (1671), p. 106.

[2] *P.W.* (see note 7), pp. 290—91.

[3] See: D. R. Oldroyd, 'Robert Hooke's methodology as exemplified in his "Discourse of Earthquakes"', *British Journal for the History of Science* **6**, 109—130 (1972); also R. Rappaport, 'Hooke on earthquakes: lectures, strategy and audience', *The British Journal for the History of Science* **19**, 129—46 (1986).

[4] A. J. Turner, 'Hooke's theory of the earth's axial displacement: some contemporary opinion', *British Journal for the History of Science* 7, 166—70 (1974).
[5] D. R. Oldroyd, 'Geological controversy in the seventeenth century: "Hooke vs Wallis" and its aftermath', in M. Hunter and S. Schaffer (eds.), *Robert Hooke: New Studies*, Woodbridge (1989), pp. 207—33.
[6] The 'hidden agenda' for this debate was apparently the question of whether preference should be given to 'naked-eye' or instrument-aided astronomical observations. Hooke was in favour of the use of astronomical instruments. He was also engaged in such a debate with the Polish astronomer, Hevelius, whose part in the controversy was taken by Wallis.
[7] R. Waller (ed.), *The Posthumous Works of Robert Hooke* ... , London (1705); hereafter *P.W.*
[8] See: Rappaport, *op. cit.* (note 3); also Y. Ito, 'Hooke's Cyclic Theory of the Earth in the Context of Seventeenth Century England', *British Journal for the History of Science* 21, 295—314 (1988).
[9] A. P. Rossiter, 'The First English Geologist', *Durham University Journal* 29, 172—81 (at 180) (1935).
[10] *P.W.*, p. 334.
[11] *P.W.*, p. 361. Hooke recommended examination of the alignments of English cathedrals, see see whether there was evidence of any systematic meridional change since their construction. Also, he hoped to find astronomical evidence for changes in the meridian. But he was not able to furnish this satisfactorily.
[12] *P.W.*, p. 372.
[13] Turner, *op. cit.* (note 4).
[14] *P.W.*, p. 374.
[15] Relating to Phœnecian settlements in Africa.
[16] The text is preserved under the title 'ΑΝΝΟΝΟΣ ΚΑΡΧΗΔΟΝΙΩΝ ΒΑΣΙΛΕΟΣ ΠΕΡΙΠΛΟΥΣ ΤΩΝ ΥΠΕΡ ΤΑΣ ΗΡΑΚΛΕΟΥΣ ΣΤΗΛΑΣ ΛΙΒΥΚΩΝ ΤΗΣ ΓΗΣ ΜΕΡΩΝ, ΟΝ ΚΑΙ ΑΝΕΘΗΚΕΝ ΕΝ ΤΩΙ ΤΟΥ ΚΡΟΝΟΥ ΤΕΜΕΝΕΙ ΔΗΛΟΥΝΤΑ ΤΑΔΕ ... ', or '[An account of] the sea-voyage of the Carthaginian king, Hanno, around the Libyan regions of the Earth beyond the Pillars of Heracles, which he also set up in the shrine of Kronos stating as follows ... ' (See: J. Blomqvist, *Hanno's Periplus with an Edition of the Text and a Translation* [Lund, (1979)], 59 and 61.) The text that has been preserved is regarded as a Greek translation of the original Punic document, certainly pre-400 B.C. and very possibly as early as 500 B.C. The edition used by Hooke consisted of an addendum to the works of Stephen of Byzantium, Leiden (1674), with a commentary by Abraham Berkeley. (He possessed a copy of this in his library.) This, Hooke mentioned, contained a commentary by the Swiss scholar, Conrad Gesner (which had been published in 1559). Hooke (*P.W.* p. 375) provided his own translation of Hanno's text.
[17] *P.W.*, p. 376.
[18] Much scholarly effort has gone to determine where Hanno did and did not go. It is possible to make reasonable geographical sense of Hanno's recorded observations without supposing that he encountered Atlantis. Hooke seems to have been mistaken about the direction of Hanno's journey, for he construed the text as saying that Hanno stayed more or less on the same latitude as Carthage, so that the lands visited would indeed have been well off the coast, perhaps near the Canary Islands. For a reconstruc-

tion of Hanno's itinerary, which holds that Hanno travelled down the west coast of Africa, see: E. Newby, *The World Atlas of Exploration*, London (1982) pp. 22—23.

[19] *P.W.*, p. 404.

[20] It is difficult to determine which passage Hooke meant. There seems to be no passage in Book 2 of Strabo's *Geography* that makes precisely the same points that Hooke mentions. There is a section in Book 1, Chapter 3.3, where Strabo recounted Eratosthenes' view that changes in the Earth take place due to the 'action of water, fire, earthquakes, volcanic eruptions and other similar agencies' (*The Geography of Strabo* ..., trans. H. L. Jones and J. R. S. Sterritt, London and Cambridge (1971—32), Vol. 1, p. 179). Strabo also mentioned (p. 175) that Eratosthenes gave credence to stories of islands beyond the Pillars of Hercules. This, however, did not give much support to Hooke: Strabo was writing in this section to criticize Eratosthenes and he held Eratosthenes' belief in 'fables' about these islands up to ridicule. Hooke's passage is more probably in Book 2, Chapter 3.6, where Strabo commends Poseidonius for citing Plato's Atlantis story and for regarding it as possibly true (*ibid.*, p. 391).

[21] Book 2, Chapter 92. Pliny actually did not give much support to Plato. (*Pliny Natural History*, trans. H. Rackham and D. E. Eichholz [Cambridge and London (1958—69)], Vol. 1, p. 335.)

[22] Pliny spoke of many earthquakes: *ibid.*, Book 2, Chapters 81—97, pp. 331—41.

[23] *P.W.*, p. 405.

[24] Strabo, *op. cit.* (note 20), Vol. 7, Book 16, Chapter 2.24, p. 269.

[25] 'On Marvellous Things Heard', *Aristotle: Minor Works*, trans. W. S. Hett, London and Cambridge (1963), Chapter 84, p. 405. Hooke's quotation is accurate, but this is a book listing strange things heard of: each bit of information begins 'they say'. Some of the information is quite plausible, other parts not so. (For instance, there is a report [Chapter 25] of mice that eat iron.) So this text was hardly a discerning choice of support; and modern scholars agree that it is not a genuine work of Aristotle.

[26] Hooke might have referred to: *The Worke of Pomponius Mela ... Translated ... by A. Golding*, London (1585).

[27] *Diodorus of Sicily ...* , trans. C. H. Oldfather *et al.*, London and Cambridge (1936—1967), Book 5, Sections 19—20, Vol. 3, pp. 145—51. The island was described as a very attractive, Atlantis-like place.

[28] A. Kircher, *Mundus Subterraneus ...* , Amsterdam (1678), Vol. 1, p. 82. Hooke owned several of Kircher's books and often quoted from them.

[29] *P.W.*, p. 394.

[30] *Ibid.*

[31] *P.W.*, p. 377. We have been unable to determine which translation of Ovid Hooke used. Though he possessed a copy of the Sandys version in his library, the texts do not correspond. Neither do the well-known translations of Golding and Dryden. (See notes 35 and 36.)

[32] *P.W.*, p. 395.

[33] Marcus Terentius Varro (116—27 B.C.) was a Roman scholar who wrote on almost every aspect of contemporary learning.

[34] For an investigation of particular interpretations of Ovid from Classical times to the Renaissance, see: 'Undermeanings in Ovid's Metamorphoses', in: D. C. Allen, *Mysteriously Meant: The Rediscovery of Pagan Symbolism and Allegorical Interpretation in the Renaissance*, Baltimore and London (1970), pp. 163—99.

[35] *The XV Books of P. Ouidius Naso, Entytuled Metamorphosis, Translated . . . by Arthur Golding* . . . , London (1567).
[36] G. Sandys, *Ovid's Metamorphosis Englished, Mythologiz'd, and Represented in Figures*, Oxford (1632), pp. 19—24.
[37] See: Allen, *op. cit.* (note 34), p. 196.
[38] *P.W.*, p. 396.
[39] *P.W.*, p. 391.
[40] *P.W.*, pp. 313—14.
[41] *P.W.*, pp. 314, 319—20.
[42] *P.W.*, p. 307.
[43] See page 146 and Oldroyd, *op. cit.* (note 3).
[44] *P.W.*, p. 422.
[45] *P.W.*, p. 427.
[46] For a history of this struggle in this period, see: F. C. Haber, *The Age of the World: Moses to Darwin*, Baltimore (1959); and D. C. Allen, *The Legend of Noah*, Urbana (1963).
[47] T. Burnet, *The Theory of the Earth* . . . , 2nd edn., London (1691).
[48] E. Warren, *Geologia: or a Discourse Concerning the Earth before the Deluge* . . . , London (1690).
[49] W. Whiston, *A New Theory of the Earth* . . . , London (1696).
[50] A. Kircher, *Arca Noe* . . . , Amsterdam (1675).
[51] M. Hale, *The Primitive Origination of Mankind* . . . , London (1677).
[52] J. Woodward, *Essay Towards a Natural History of the Earth* . . . , London (1695). Hooke rejected the idea that the inland location of the figured stones could be accounted for by the Flood, for not nearly enough time was involved. *P.W.*, p. 341 and *passim*; most clearly p. 412.
[53] *P.W.*, p. 407. The reference is accurate, but the interpretation doubtful: *Herodotus* . . . , trans. A. D. Godley, London and Cambridge (1926), Book 2, Chapter 12, p. 287.
[54] Perhaps *Meteorologica*, Book 1, Chapter 14. Here Aristotle gave Egypt as an example of land being built up through the deposits of rivers; but Hooke's figures of the rate of deposit were not mentioned. See: Aristotle: *Meterologica*, trans. H. D. P. Lee, London, and Cambridge, (1962), p. 111. *P.W.*, p. 407.
[55] *Ibid.*, Book 1, Chapter 14, p. 119. *P.W.*, p. 411.
[56] *Metamorphoses*, Verses 262 ff. *P.W.*, p. 410.
[57] Allen, *op. cit.* (note 34), p. 58.
[58] Strabo, *op. cit.* (note 20), Book 10, Chapter 3.23, p. 119.
[59] For a discussion of euhemerism in this period see: J. D. Cooke, 'Euhemerism: A mediæval interpretation of classical paganism' *Speculum* 2, 396—410 (1927).
[60] W. Ralegh, *The History of the World* . . . , London (1614).
[61] F. Bacon, *Of the Wisdom of the Ancients* in: J. M. Robertson (ed.), *The Philosophical Works of Francis Bacon*, London (1905), p. 822.
[62] *Ibid.*, p. 827.
[63] Bacon, *The Proficience and Advancement of Learning*, Book 2, in *Philosophical Works, op. cit.* (note 61), p. 89.
[64] H. Reynolds, *Mythomystes Wherin a Short Survey is Taken of the Nature and Valve of Trve Poesy* . . . , London (1632).

[65] T. Sprat, *The History of the Royal Society of London*..., London (1667), p. 414.

[66] The tradition of physico-mythology initiated by Hooke — the idea that myth somehow hides true historical information which may give support to 'modern science or natural philosophy — has persisted. Interesting modern versions appear in Georgio de Santillana and Hertha von Dechend's *Hamlet's Mill: An Essay on Myth and the Frame of Time*, Boston (1977); Immanuel Velikovsky's now discredited *Worlds in Collision*, London (1950) and *Earth in Upheaval*, London (1956); and the more reliable D. B. Vitaliano's *Legends of the Earth: Their Geologic Origins*, Secaucus (1976).

[67] *P.W.*, p. 334.

[68] N. Copernicus, *On the Revolutions of the Heavenly Spheres*, trans. C. C. Wallis, Chicago (1952), pp. 508 and 515.

[69] S. Shapin, 'The house of experiment in seventeenth-century England', *Isis* **79**, 373–404 (1988).

[70] See: P. Dear, '*Totius in verba*: rhetoric and authority in the early Royal Society', *Isis* **76**, 145–61 (1985).

[71] This is based upon the published catalogue for the sale of Hooke's personal library: R. Smith (compiler), *Bibliotecha Hookiana*, London (1703) (with appendix). The classification we offer is necessarily approximate, since some of the titles and authors cannot be identified satisfactorily. Smith's lists are given in highly abbreviated form and many of the books cover more than one topic. Nevertheless, the lists give a good idea of Hooke's tastes and interests. The books were in English, Latin, French, Spanish or Italian, with just a few Dutch texts. The compiler informed the public in an introductory paragraph that 'the World may be satisfied that he [Hooke] was not an Idle Possessor of them [the books], [since] he hath left behind him many curious notes on some, considerable M.S.S. Improvements to others, not unworthy the View and Perusal of the Virtuosi of the Age ...'. Hooke's technical works were chiefly in Latin or English, whereas his 'lighter' reading tended to be in French, Spanish or Italian. It would appear that he didn't read German.

[72] These include, broadly, works of classical authors such as might be studied by classics students today, *including* histories. We classify elsewhere works such as Ptolemy's *Almagest*, which Hooke would have purchased for their technical interest.

[73] This includes works that are concerned with more than one topic, or texts that we have been unable to identify.

[74] The view attributed to Fabianus was that, were a deluge to occur, the whole universe would be shattered. Seneca supposed that such a catastrophe would have no single cause, but all the forces of nature were working to cause the seas to rush violently from their places. This, however, would be because the seas could no longer hold their boundaries due to the influx of water from rivers, not because of earthquakes. Seneca had the idea that catastrophes recurred as the world passed through cycles of destruction and re-birth. Servius Flavius Papirius Fabianus lived in the first century A.D. and wrote a work entitled *Natural Causes*.

[75] Ovid, *Metamorphoses*, trans A. D. Melville, Oxford (1986), p. 5. (Hereafter *M*.)

[76] In fact, the word '*Lucaon*' means 'wolf-skin'; and '*Lyois*' means dissolution.

[77] *P.W.*, p. 414.

JOHN GASCOIGNE

'THE WISDOM OF THE EGYPTIANS' AND THE SECULARISATION OF HISTORY IN THE AGE OF NEWTON

One of the basic tensions with which the proponents of the Scientific Revolution had continually to contend was to find some accord between the mechanical philosophy — which was found to provide every-increasing explanatory utility — and the concept of Providence, which underlay the religious foundations of their society. If, as the mechanical philosophy appeared at times to suggest, all was determined by particles in motion it was difficult to find a place for a guiding Providence impressing his will on Creation; or, at best, the Deity was consigned to a marginal role as a remote initiator who thereafter took little interest in terrestrial affairs. But in a society in which religion was inseparably intertwined with the political and social order it was essential to find some sort of a *modus vivendi* between the mechanical philosophy, which tended towards a form of determinism, and a conception of the role of Providence which still allowed for God's involvement in directing human affairs. That the scientific developments to which we give the name the Scientific Revolution continued to grow and prosper — despite such setbacks as Galileo's condemnation — is an indication that early modern European society was persuaded that such a reconciliation was possible.

The Scientific Revolution, then, involved not only a transformation in humankind's understanding of the physical universe but also a reappraisal of the theological conceptions which underlay so many facets of early modern society. As Funkenstein[1] has recently argued at length, the intellectual matrix in which the Scientific Revolution was formed was the result in large part of a long process of theological and philosophical debate, which had its roots in the Middle Ages, out of which developed an understanding of the role of Providence which allowed for the possibility of uniform physical laws. Such a conception of God's relations with the physical world was to underlie the mental universe of many of the major figures we associate with the Scientific Revolution — in particular Kepler, Boyle and Newton regarded the lawlike and mathematically predictable behaviour they discerned in nature as instances of the mind of the Creator at work in shaping and sustaining his creation. By the late seventeenth century, then, natural philosophers and, to an increasing extent, theologians, tended to emphasise the regularity and predictability of Providence's actions. In the language of the theologians ordinary (or general) Providence — God's direction of the universe as a whole through a set of uniform laws — received greater emphasis than extraordinary (or particular) Providences — occasions on which God intervened directly in human or terrestrial affairs. Or, to put the same intellectual transformation in the language of the natural philosophers, secondary causes — the instruments of God's will — tended to be given closer attention than the final causes which were concerned with God's ultimate purposes.

One further reflection of this same change in mentality was the growing tendency to regard the study of history, as well as the study of nature, as principally concerned with understanding those natural, secondary causes which were Providence's agents rather than with directly tracing God's role in human affairs. As in natural philosophy so, too, in history there was an increasing reluctance to explain events by involving God's direct intervention; while few historians (or natural philosophers) denied God's ultimate control of events Providence was more and more portrayed as working through the natural processes of history. Such an understanding of history — which we may, for convenience, describe as more secular (though, not necessarily, anti-religious) in character — was not altogether new in the seventeenth century. The Renaissance humanists had drawn on the example of classical historians to promote the study of history as a discipline with its own methods and concerns which could be distinguished from those of theology.[2] But, as Smith Fussner and Sypher have argued, such an approach to history may well have been strengthened by the increasing tendency of the Scientific Revolution to separate the study of events explicable in terms of secondary causes from the final causes which were the natural concern of the theologians.[3] Certainly the realisation of such an intellectual demarcation was a lifelong goal of Francis Bacon and was reflected in his historical as well as his scientific writings. It is the argument of this essay that such a secularised approach to historical issues had, by the late seventeenth century, become so pervasive that it coloured not only historical accounts of political events, where the influence of classical models and the Renaissance humanists was naturally strong, but also shaped the way in which the biblical record of the history of the Chosen People was understood. Such a reappraisal of the role of Providence in relation to so sensitive an area as the interpretation of Scripture was, moreover, to be closely linked with Newton and a number of his contemporaries who sought to defend true religion both from the superstitious and the unbelieving by showing how Providence worked primarily through secondary causes both in nature and in history.

The basic outlines of a form of Christian historical apologetics had been developed in the early Church in the context of the debates between Christian and pagan scholars. Some Church Fathers and, in particular, the influential Clement of Alexandria (d. 213 AD) had argued that classical learning, including that of the Greeks, had been largely derived from the Jews — a line of argument which, not surprisingly, had previously been advanced by Hellenistic Jews.[4] The writings of the supposed ancient Egyptian seer, Hermes Trismegistus, were used both by the Church Fathers and, subsequently, by Renaissance scholars such as Ficino to defend an extension of this position: that the Greeks not only derived their learning directly from the Jews but also indirectly through the Egyptians who, in turn, had been greatly influenced by the Jews.[5]

This ancient problem of the relations between the history of the Jews and other peoples of the ancient world became a matter of renewed concern after the Renaissance when the recovery of an increasing number of classical texts focussed attention on the problem of reconciling the biblical chronology with that which could be derived from pagan authors. Much scholarly labour was devoted to harmonising the biblical chronology with the classical histories of Greece and Rome and with the account by Greek historians of the Egyptians and other ancient civilizations of the Middle East. Newton's chronological writings formed part of this erudite tradition and, like many of his scholarly brethren in the field, he developed ingenious explanations to help to explain why the claims of Egypt to a past much older than that of the Jews could be dismissed — thus Newton argued that the supposed ancient pharoah, Sesostris, was but another name for Sesac, who posed no challenge to the biblical chronology since he was recorded in the Old Testament as invading Judea after the death of Solomon.[6]

Underlying the debate about such arcane matters of chronology was the desire by Christian scholars to defend a conception of world history which accorded primary importance to the Jews. In a society in which the biblical account still provided the basic framework for understanding history a defence of the historical primacy of the Jews as against the Egyptians, or other Gentile peoples, was also a means of defending an essentially providentialist conception of history in which God's relations with his Chosen People provided history with its most ancient and enduring theme. To further bolster this providentialist conception of history, with its premise that the biblical account of the Jews provided the key for understanding the history of the ancient world, a number of late seventeenth-century scholars restated the traditional argument that the Jews were not only the most ancient people but also that it was from them that the learning and even the languages of the ancient world were derived.

One of the lengthiest defences of this view was the work of Theophilus Gale, a dissenting divine, whose vast compilation of erudition, *The Court of the Gentiles* (1669—77) set out to demonstrate

That the wisest of the Heathens stole their choicest Notions and Contemplations both Philologic, and Philosophic, as wel Natural and Moral, as Divine, from the sacred Oracles.

One chapter of the work was even devoted to the theme 'Of the Traduction of al[l] Languages and Letters from the Hebrew'. The origin of this dissemination of Jewish learning to the Egyptians and hence to the Greeks Gale traced back to the period when Joseph and other partriarchs dwelled in Egypt for, before then, 'the Egyptians were no way famous for Wisdom, or Philosophie'. Much later, the Egyptians and the Greeks also benefited from the translation of the Hebrew Bible into Greek.[7] Among Gale's readers was the mathematician, John Wallis, who invoked Gale's authority when arguing that:

'Tis well known (to those conversant in such Studies) that much of the Heathen Learning (their Philosophy, Theology, and Mythology) was borrowed from the Jews, though much Disguised, and sometimes Ridiculed by them.[8]

Though the late seventeenth century saw the confident restatement of this ancient form of apologetics it was also in this period that a number of scholars began to question the traditional importance accorded to the Jews. And, though this scholarly re-evaluation was largely the work of those within the Christian fold, this tendency to downplay the significance of the Jews brought with it the need to re-evaluate the highly providentialist understanding of world history which for so long had been associated with the view that the biblical account of the Chosen People was the lynch-pin for any understanding of the history of the ancient world. But such a reappraisal of world history did not necessarily mean abandoning a providentalist conception of history altogether — just as Christian exponents of the mechanical philosophy argued for the importance of understanding the way in which God used secondary causes to achieve his ends so, too, those engaged in developing a more critical understanding of the relations of the Jews with other peoples and cultures were frequently to argue that their work was an illustration of the way in which Providence could work through the normal processes of human history. The God of such historians, as of many of the mechanical philosophers, was one who intervened in terrestrial affairs only on rare occasions but one, nonetheless, whose guiding hand could be discerned in the overall pattern of events.

Such an understanding of both the workings of history and of nature can be seen in the writings of John Spencer, whose work did much to prompt a re-evaluation of the traditional conception of the relations between the Jews and their neighbours in the ancient world. While master of Corpus Christi College, Cambridge (a post he held from 1667 until his death in 1693), Spencer produced his pioneering work in the field of comparative religion, the massive *De Legibus Hebraeorum Earum Rationibus* . . . (1685). In it he explored in great detail and with formidable erudition the theme he had already ventured upon in his relatively brief, *Dissertatio De Urim and Thummim* (1669) — a discussion of the origin of Hebrew divinatory practices — namely the extent to which the Jews derived their religious practices from the Egyptians. Spencer's answer was that, although some of the Jewish laws were framed in opposition to Egyptians and other Gentile practices, nonetheless

God tolerated and transferred not a few of the rites that were in use among the Pagans into his own law and worship; after he had corrected and reformed them.[9]

Spencer pointed out that the Egyptians of the age of Moses were a deeply conservative people who were unlikely to borrow rituals and customs from others — least of all from the Jews whom they despised. Moreover, the Greeks, who freely acknowledged their debt to the Egyptians, had nothing to say about the Jewish origins of Egyptian culture. Thus the close parallels between Jewish and Egyptian customs which Spencer minutely described in relation to a whole range of practices from the cult of worshipping God in a tabernacle to funeral customs could only mean that the Jews derived such practices from the Egyptians — just as the Egyptians had earlier been influenced by the customs of such neighbouring peoples as the Syrians.

Such a view ran counter to the commonly accepted Christian framework of world history in which the Jews were portrayed as the fountain head of the civilization of the ancient world. But in Spencer's view such a revision of the received wisdom did not entail any breach with orthodoxy. For it was Spencer's goal to show how God used the natural processes of history and, in particular, the cultural assimilation of Egyptian practices among the Jews to win over his Chosen People. Thus he stressed that the Jews did not come to a knowledge of the true God as a result of a single, dramatic, divine intervention in human affairs but rather that it was the result of a slow, often imperfect process of development. The ancient Jews, he wrote were 'rude, ignorant, obstinate' and very 'superstitious' so that

> it was almost necessary that God should indulge them the use of some of the ancient [Gentile] rites, and accommodate his sanctions to their taste and capacity.

In particular God allowed the Jews to retain

> very many of the Egyptian usages; especially those, that by their pompous appearance and shew were likely to please and take with the populace.[10]

Spencer, then, saw his work as contributing to an understanding of the way in which Providence could use secondary causes to influence history. True religion had to work through the natural limitations of human history — even the distinctive rituals of Christianity took some time to develop from their Jewish origins, a process that, by implication, Spencer parallels with the way in which Judaism gradually developed from earlier forms which owed much to the Gentile cultures of the Near East.[11] In Spencer's understanding of history, then, God does not achieve his goals 'immediately and as if through straight lines'[12] but rather is generally content to shape human affairs by the natural process of history with all its attendant delays and complications.

A similar conception of the way in which Providence works through secondary causes, rather than by means of direct intervention, can also be found in Spencer's work dealing with God's relations to the natural world. This theme he developed in a Cambridge sermon entitled *A Discourse Concerning Prodigies* . . . which was delivered in 1663 in the still tense aftermath of the Restoration when many feared that religious enthusiasts could undermine the restored order in Church and State.[13] Thus it was Spencer's purpose to show that the comet of that year, and indeed all such prodigies, should not be seen as harbingers of providential intervention. Such an aim, argued Spencer, was not only 'profitable to serve the just interest of the State' but also 'serve[d] the honour of Religion which the common reverence of Prodigies doth greatly trespass upon'.[14] For the argument which underlies Spencer's sermon is that God's will is manifest in the regular, law-like behaviour of the universe and that occasional spectacular events (such as comets) still conform to such divinely ordained laws; in short, he emphasizes the role of ordinary rather then extraordinary Providence. Thus he sees it as a matter of religious awe that one could discern

the faithfulness of Nature to its Original Laws of motion, the continuance *of all things as they were from the beginning of the Creation.*

God, he continues, with a metaphor natural to an author,

never saw it necessary (as upon maturer thoughts) to correct and amend any thing in this great Volume of the Creation since the first edition thereof.

Thus God's will is achieved not through the sort of direct intervention which some claimed to see in the case of comets but rather through his control of 'the several motions and mutual aspects of Secondary Agents, from the beginning of time to the end thereof'.[15] In support of his view that since comets and other such prodigies could be explained as the manifestations of a set of uniform cosmic laws they do not constitute a direct providential intervention in human affairs he invokes the authority of Kepler and Tycho Brahe. Spencer uses their work to demonstrate both that the traditional Aristotelian understanding of comets as a fiery 'exhalation' of the sub-lunar world was false and that the distance of comets from the earth undermined the view that terrestrial evil was the result of 'the malign aspects of Comets'.[16]

Though Spencer's basic aim was to 'assert *Prodigies* to be none of God's designed tokens' he was mindful, nonetheless, of the delicate intellectual balancing act that such a line of argument required of him as an Anglican opponent not only of enthusiasts but also of unbelievers. Spencer was well aware that while too great an emphasis on the role of direct providential intervention might lead to superstition and enthusiasm an exclusive concern with the natural means by which God's will is achieved might lead to infidelity. As Spencer himself put it:

> As we must not loose [sic] our Philosophy in Religion, by a total neglect of second causes and turning superstitious; so neither must we loose our Religion in Philosophy, by dwelling on second Causes, till we quite forget the *First* and become profane.[17]

Spencer's solution was to emphasise the ultimate dependence of such secondary causes on the will of God — a point he made in the manner of a true mechanical philosopher by invoking the analogy between the world and a clock which

> though it contain[s] very strong and powerful Springs of action within it self ... yet these blind and decaying Powers must be managed and perpetually wound up by an Hand of Power and Counsel.[18]

This balance between the role of first and secondary causes was one which Spencer sought to achieve not only in his natural philosophy through his emphasis on the role of Providence in creating and sustaining a system of uniform laws but also in his historical writings by arguing that God used secondary causes — such as the cultural influence of the Egyptians — to help bring into being the form of religion which was to develop into the mature Judaism of God's Chosen People.

Predictably, however, Spencer's *De Legibus*, with its implied critique of the primary place traditionally accorded to the Jews in the history of the ancient world prompted considerable controversy. Spencer's view of Providence as employing the natural processes of human history to bring to fulfillment the Jewish religion prompted Hermann Witsius, a professor at Utrecht, to decry in his work, *Aegyptiaca* (1683), such an interpretation since

> it is a dishonouring of GOD, who has the hearts of men in his power ... to conceive of him as standing in need to the tricks of crafty politicians.[19]

Closer to home, John Edwards, another Cambridge don, also reacted sharply against Spencer's attempt to demonstrate the way in which God shaped history through the natural processes of human society. Edwards, a vigilant defender of Christian orthodoxy who was to be one of Locke's most pertinacious critics, devoted part of his modestly entitled *Polgpioikilos Sophia. A Compleat History or Survey of All the Dispensations and Methods of Religion, From the Beginning of the*

World to the Consummation of All Things ... (1699) to arguing that the way in which Spencer portrayed the Jews developing many of their religious rituals from forms derived from the Egyptians and other Gentile peoples made 'the True God most diligently and precisely tread in the steps of the false Gods and Idols'. Edwards then vigorously reasserted the traditional primacy accorded to the Jews by setting out to demonstrate

that many of the Pagan Rites and Customs in Religion (as well as in Secular Affairs) were borrow'd from the *Jews* and their Secular Usages

— something, he added, with a pointed reference to Spencer,

which is directly contrary to what this author asserts, *viz* that the Rites and Ceremonies injoyn'd by God himself to the *Jews* were of Pagan Extraction.[20]

In an earlier work, *Cometomantia. A Discourse of Comets* ... (1684) Edwards had attacked another aspect of Spencer's portrayal of the role of Providence: his argument that comets were not portents of calamity since God's will was made manifest in nature through uniform laws rather than divine intervention. Just as Edwards saw Spencer's conception of history as undermining the role of an active Providence so, too, he saw Spencer's form of natural philosophy as tending the same way. For Edwards 'A Christian Divine cannot have a better Text to preach Repentance from than a Comet' since a comet was an overt sign of the immediate and direct involvement of Providence in human affairs. God, he argued, 'speaks to and instructs Mankind' not only through Scripture but also 'by the course of Nature, and by Acts of Providence'. Edwards concedes that the Moderns had shown that some comets were above the moon but maintained that at least some others were sublunar, as Aristotle had taught, and could therefore influence human affairs.[21] Edwards' immediate target was probably Pierre Bayle whose famous pamphlets on comets — in which he, like Spencer, argued that the appearance of a comet had no supernatural significance — were published in 1682 and 1683,[22] just before Edwards' work. However, as a fellow Cambridge resident Edwards must have known Spencer's influential work on prodigies and there is little doubt that, along with other clerical adversaries such as Gassendi, Edwards had Spencer in mind when he castigated those among the clergy who undermined belief in comets as divine portents. Such writers, stormed Edwards:

carry on the Plot of Atheists and Epicures to root out the Notion of a God, to extirpate Providence, to debauch Mens Lives and Manners, and to blot out the Sense of another World.[23]

The geologist, physician and lay Christian apologist, John Woodward, expressed his reservations about Spencer's historical work more temperately than the irascible Edwards and acknowledged his 'infinite industry' but he, too, regarded Spencer's challenge to the commonly accepted framework of world history as a threat to orthodoxy. Though he acknowledged that it may have been unintentional Woodward viewed Spencer's work as tending in the same direction as Giordano Bruno 'and others of like libertine principles, who bear no good-will to Christianity' and who, in order to subvert it, 'extoll the Egyptians beyond measure, decry the Jews, and vilify that nation, their archives and laws, as meerly of Egyptian extract'.[24] Woodward's response in his posthumously published *Of the Wisdom of the Ancient Egyptians* was to undermine such an historical interpretation by a systematic denigration of Egyptian civilization and learning. The Egyptians, he wrote 'were the most ostentatious, boasting people in the universe' and such much vaunted achievements as the pyramids were vastly overrated. Far from being the source of many of the Jewish rituals the Egyptian religion

> was undoubtedly the wildest and most fantastic that the sun ever saw ... They were above all other nations, so sunk in idolatory, that they seem to have known little if anything of God.

Indeed, Spencer's whole thesis needed to be turned on its head — it was the Egyptians that derived many of their rituals and laws from the Jews rather than vice versa.[25]

Along with Spencer, another of Woodward's targets in this same work was Thomas Burnet who, in his *Archaeologiae Philosophicae sive Doctrina Antiqua de Rerum Originibus* (1692), had, like Spencer, contrasted the civilization of the Egyptians with the backwardness of the Jews. Burnet had in fact gone so far as to suggest that the Mosaic cosmogony was (as Woodward summarises him)

> drawn up ... in imitation of those of their neighbours the Egyptians, Phoenicians and Chaldeans, all of whom the Jews want greatly to admire.[26]

Burnet's *Archaeologiae Philosophicae*, then, complemented his earlier, much better-known work, *The Sacred Theory of the Earth* (1681),[27] in which he had argued that the biblical account of the creation of the earth should largely be understood metaphorically. Burnet had then devoted the bulk of his *Sacred Theory* to demonstrating that the formation of the world could largely be explained in terms of natural forces even though the scriptural record could be used to supplement and clarify those aspects of the world's history which were otherwise obscure. As Burnet himself wrote:

> This theory being Chiefly Philosophical, Reason is to be our first Guide; and where that falls short or any other just occasion offers itself, we may receive further light and confirmation from the Sacred writings.[28]

Burnet was reluctant, however, to invoke miracles or supernatural explanations too readily for the whole thrust of his work was to emphasise the extent to which God worked through the natural processes of nature by means of his ordinary rather than his extraordinary Providence; or, as Burnet himself put it, his aim was to bring out

> the true Notion and State of *Natural Providence*, which seems to have been hitherto very much neglected, or little understand in the World.

It followed, then, that

> if we would have a fair view and right apprehensions of Natural Providence, we must not cut the chains of it too short, by having recourse, without necessity, either to the First Cause, in explaining the Origins of things: or to Miracles, in explaining particular effects. This, I say, breaks the chains of Natural Providence ... Neither is anything gain'd by it to God Almighty; for 'tis but ... to take so much from his ordinary Providence, and place it to his extraordinary.[29]

It was no accident that Woodward produced not only a critique of Burnet's (and Spencer's) historical views but also a much more influential attack on his cosmogony. For, in Woodward's view, both Burnet's historical and cosmogonical systems served, however unintentionally, to weaken the foundations of true religion since both greatly reduced the extent of providential intervention in terrestrial affairs. Thus it was the aim of Woodward's *Essay towards a Natural History of the Earth* (1695) to establish that the form of the terraqueous world could only be explained by God's direct intervention as explicated in Scripture — an endeavour which predictably was enthusiastically received by John Edwards who saw it as confirming that 'what we behold in the world is a Proof of a Deity and Providence'.[30] Woodward placed particular emphasis on the significance of the biblical Deluge which he viewed as necessitating a virtual second Creation which could only be explained in miraculous terms. In explicit opposition to Burnet he affirmed that

the Deluge did not happen from an accidental Concourse of Natural Causes, as the above-cited [Burnet] is of Opinion; that very many Things were then certainly done, which never possibly could have been done without the Assistance of a *Supernatural Power.*[31]

Thus Woodward saw his account of the Deluge as demonstrating 'the Fidelity and Exactness of the Mosaic Narrative of the Creation'.[32]

Just as Woodward's emphasis on direct providential intervention was reflected both in his writings on cosmogony and history — the latter of which argued at length for the overriding historical importance of the Jews as God's Chosen People and the consequent cultural insignificance of the Egyptians — so, too, Burnet's historical work, like his cosmogony, argued for a more naturalistic interpretation of the role of Providence in history. Even in his *Sacred Theory* Burnet had implied agreement with the general tenor of Spencer's historical views by commenting that 'as to the *Jews*, 'tis well known that they have no ancient Learning, unless by way of Tradition, amongst them.'[33]

This point he developed further in his *Archaelogiae Philosophicae* (1692) in which he argued that the Mosaic law was influenced not only by the Egyptians but also by another Gentile people, the Zabians. In considering the 'Contention' between the Jews and the Egyptians 'about the Precedency and Antiquity in learned Discoveries' he sided with the Egyptians on the grounds that the Greeks freely acknowledged their intellectual debt to the Egyptians but made no mention of the Jews. Moreover, against those who viewed Moses as the source of all ancient learning he argued that

it appears from the sacred Scriptures, that the *Egyptian* Wisdom was more ancient than his [Moses'] and he was the Disciple rather than the Teacher of that learned Nation.

As for those who would go further back and claim Abraham as the source of Egyptian learning Burnet urged them to consider how short a time Abraham was in Egypt and how unlikely it was that 'a Stranger, of a different Language, [could] in less than two Years Space, instruct a Nation'. But Burnet did not altogether reject the biblical historical framework since he argued that 'the Original of the Barbaric Philosophy' could be traced back to 'the Deluge, and Noah the common Father of *Jews* and *Gentiles*'. For Noah as an 'Inhabitant of both Worlds' —

the ante-diluvian and post-diluvian — 'delivered the Lamp of Learning from one to the other' even though, with the passage of time, this learning 'very much declined, and ... those seminal Doctrines were almost choaked by the prevailing Tares'. Indeed, the restoration of such ancient learning through 'the Principles of Nature and clear Reason' Burnet saw as paving the way for the coming of the millenium for

when the End of all Things approaches, Truth, being revived, may shine with double Lustre, as the Prelude of a future Renovation.[34]

It followed from such a view of world history that aspects of the true original Noachian religion and learning could be found in the different cultures of antiquity whether Jewish or Gentile — that there was not, in short, the great gulf between the civilization or even the religion of the Jews as God's Chosen People and that of their pagan neighbours which had been fundamental to the traditional understanding of the way in which God had shaped world history. This same tendency to downplay direct providential intervention in human affairs can also be seen in the way in which Burnet, both in the *Sacred Theory* and the *Archaeologiae Philosophicae*, attempted to demythologise aspects of the biblical record. In defending his departure from the Mosaic account of Creation in the former he argued that

if we should follow the Vulgar Style and literal sense of Scripture ... we must renounce Philosophy and Natural Experience ... [but] Scripture never undertook, nor was ever designed to teach us Philosophy, or the Arts and Sciences.[35]

But in the *Archaeologiae Philosophicae*, which he described as a 'Commentary, or Appendix to the *Theory of the Earth*,'[36] he took this principle further to imply that Scripture was not only a partial guide to natural philosophy but also to history. Thus he argued that the biblical account of the Fall in the Garden of Eden included 'some things Parabolical, and, which will not bear a construction altogether Literal'. He questions for example how Eve could have been made out of a rib and how a serpent could speak.[37] Hence a popular jingle of the time satirised Burnet and others for maintaining that

All the Books of Moses
Were nothing but supposes.[38]

Nonetheless, Burnet's protestation in his *Archaeologiae Philosophicae* that 'I call God to *Witness* that in this or any other Writing I never proposed more to my self, than the Promotion of Piety founded upon Truth'[39] was no doubt sincere. For Burnet saw himself as distinguishing the essential religious message of the Old Testament, which was still relevant to his own enlightened age, from the poetic setting in which it had been set to meet the needs of a primitive people.[40] Moses, he wrote, had 'followed the popular System; that most pleases the People ... And in so doing, he rightly consulted the Publick Safety.'[41] Not surprisingly, however, the *Archaeologiae Philosophicae* brought Burnet's clerical career to an abrupt halt. The eighteenth-century historian, John Oldmixon, reported that on the death of John Tillotson, the Archbishop of Canterbury, in 1694 Burnet's name was put forward as a possible successor

with some Prospect of success, till upon a Representation of Certain Bishops, that some of his Writings were too Sceptical, another Divine [Tenison] was pitch'd upon.[42]

However, Burnet was left undisturbed as master of Charterhouse [school] — an indication of the theological latitude that could exist in the Established Church, particularly after the revolution of 1688.

Burnet, like Spencer, had been at Cambridge (where he was a fellow of Christ's College before he left to become master of the Charterhouse in 1685) at the same time as Newton — indeed when Newton took his MA in 1668 it was Burnet who officiated as Senior Proctor.[43] Burnet and Newton had obviously become acquainted at Cambridge since, some time before publishing his *Sacred Theory*, Burnet consulted Newton who, in January 1680/1, sent a lengthy and generally sympathetic response to Burnet's overall line of argument. In particular Newton largely endorsed the form of scriptural interpretation which underlay the work: like Burnet, Newton believed that where necessary the miraculous elements of the Mosaic account of Creation should be reinterpreted since Moses had been writing for a primitive, pre-scientific people. As Newton himself put it:

As to Moses I do not think his description of creation either Philosophical or feigned, but that he described realities in a language artificially adapted to the sense of ye vulgar ... his business being not to correct the vulgar notions in matters philosophical, but to adapt a description of ye creation as handsomely as he could to ye sense & capacity of ye vulgar.[44]

Newton also concurred with Burnet's underlying belief in the importance of what he called 'Natural [or General] Providence' as against the traditional readiness to invoke God's extraordinary Providence — a sentiment pithily summarised in Burnet's italicised remark that *'The Course of Nature is Truly the will of God'*.[45] For Newton also argued that God's creative power could best be understood as working through the normal processes of nature even though he (like Burnet) acknowledged that there were rare occasions — such as at the original Creation — when God did intervene directly in the natural realm. As he wrote in his letter to Burnet:

Where natural causes are at hand God uses them as instruments in his works, but I do not think them alone sufficient for ye creation.[46]

Newton's historical writings also have much in common with those of Burnet and it is possible that the two men discussed matters historical as well as issues relating to natural philosophy. Newton's copy of Burnet's *Archaeologiae Philosophicae* indicates close study[47] although by the time that this work was published in 1692 Newton had already developed the general outlines of his historico-theological system.[48] Like Burnet, Newton regarded Noah, rather than Abraham or Moses, as the original source of true religion and learning; consequently, he, too, argued that vestiges of truth could be found among the ancient Gentile peoples as well as that of the Jews since all were descendants of Noah and his sons. Both also shared the belief that modern philosophy was contributing to the recovery of the ancient truths which had been distorted after Noah's death.

As Westfall has shown in his important article[49] on Newton's major (though unpublished) theologico-historical work, the *Theologiae Gentilis Origenes Philosophicae* Newton conceived of the true religion as that which most closely resembled that which prevailed at the time of Noah immediately after the Deluge, before the idolatry — which to Newton was the root of all evil not only in religion but also in politics and even philosophy — began to corrupt it. The subsequent religious history of humankind, as outlined in the Old and New Testaments, was, then, in Newton's views, a record of the attempts to purge religion of its idolatrous accretions. The nearest that Newton came to acknowledging publicly such a view of religious history was in a passage of his *Chronology of the Ancient Kingdoms Amended*, a work which he was preparing for publication when he died in 1727 and which was published by his executors in the following year.[50] In it he described the religion of Moses and the prophets as being based on

The precepts of the sons of Noah, which was the primitive religion of both Jews and

Christians, and ought to be the standing religion of all nations, it being for the honour of God, and good of mankind.[51]

In the privacy of his unpublished manuscripts Newton made it evident that he saw the task of restoring the true Noachian religion as continuing up to the Protestant Reformation and beyond since Newton took the view that even Protestant Christianity needed to be purged of what he considered its idolatrous trinitarianism.

Such a reading of human history, then, largely reduced the role of the Old Testament prophets, and even that of Christ, to that of religious reformers rather than the instruments of a wholly new divine revelation — for, to Newton, the true religion had been established long before either Christ or the prophets. In one of his jottings he came close to acknowledging explicitly such a considerable breach with orthodoxy for he noted as the theme of a proposed (but never written) chapter of a version of the *Theologiae Gentilis*:

What the true religion of the children of Noah was before it began to be corrupted by the worship of false Gods. And that the Christian religion was not more true and did not become less corrupted.[52]

Elsewhere, in a manuscript entitled 'A Short Scheme of the True Religion', Newton again hinted at his belief that Christianity did not constitute a wholly new revelation since residues of the true Noachian religion — or at least the ethical teaching which was to Newton the most important part of religion — could be found in cultures such as that of China which were far removed from Judaeo-Christian civilization. The principle that we should love our neighbours as ourselves, he wrote,

was the Ethics, or good manners, taught the first ages by Noah and his sons by their seven precepts, the heathens by Socrates, Confucius and other philosophers, the Israelites by Moses and the Prophets and the Christians more fully by Christ and his Apostles.[53]

Like Burnet, then, Newton departed from the traditional belief that there was in antiquity an unbridgeable gulf between the religion of the Jews and that of the Gentiles. However, Newton went considerably further than Burnet in downplaying the significance of Christ, for in Burnet's theologico-historical scheme 'the whole Hinge of Providence with respect to human Affairs, turns upon the Mystery of the Messiah'.[54]

Newton's conception of world history, then, reduced the role of Providence to intervening occasionally and, generally, through natural processes to re-establish the original true religion rather than to establish a wholly new religious order. Similarly, Newton's conception of the role of Providence in the natural order (like that of Spencer) was to establish a natural harmony at the beginning and to intervene only occasionally and, generally speaking, through natural agencies, to maintain the clock-like order established at Creation. As Newton remarked in the thirty-first (and last) query of his *Opticks* all material things were 'variously associated in the first creation by the counsel of an intelligent agent. For it became Him who created them to set them in order'. But, he continued, over time 'some inconsiderable irregularities' do arise

from the mutual actions of comets and planets upon one another, and which will be apt to increase, till this system wants a reformation.[55]

Such a 'reformation' could, of course, only be the result of providential intervention but in Newton's view, the Creator generally achieved these ends by working through secondary causes, particularly the comets.[56]

The same query (and the *Opticks* as a whole) concludes with a rather oblique reference to the theologico-historical *schema* that Newton had developed in his *Theologiae Gentilis*. As natural philosophy is perfected, he wrote, so too 'the bounds of moral philosophy will be also enlarged'[57] — this, in effect, meant a growth in true religion since for Newton true ethics, which included our duty to God, constituted virtually the sum and substance of true religion. Thus in his unpublished *Irenicum* he defined the true religion of Noah as 'The religion of loving God and our neighbour'.[58] To Newton, then, the true natural philosophy was inextricably linked with the true religion since, as he wrote in the same section of the *Opticks*:

For so far as we can know by natural philosophy what is the First Cause, what power He has over us, and what benefits we receive from Him, so far our duty towards Him, as well as that towards one another, will appear to us by the light of Nature.

Conversely, the loss of the true natural philosophy was therefore closely associated with the undermining of true religion and ethics, a loss that Newton saw as the outcome of what for him was the true Fall of Man: the growth of idolatry. Thus he continues the above passage by concluding the *Opticks* as follows:

And no doubt, if the worship of false gods had not blinded the heathen, their moral philosophy would have gone farther than to the four cardinal virtues; and instead of teaching the transmigration of souls, and to worship the Sun and Moon, and dead heroes, they would have taught us to worship our true Author and Benefactor, as their ancestors did under the government of Noah and his sons before they corrupted themselves.[59]

One of the major purposes of the *Theologiae Gentilis* was to trace the tragic history of the way in which the corruption of true religion by idolatry was accompanied by the corruption of true natural philosophy. For, as Newton's conclusion to the *Opticks* suggests, not only did he maintain that the religion of Noah and his sons was pure and undefiled so, too, was their understanding of the natural world. The central rite of the Noachian religion was the maintenance of a perpetual vestal flame at the centre of some form of shrine — what Newton called a 'prytaneum'. The prytaneum was a model of God's creation, the universe, and the fire a symbol of the sun and hence an affirmation of a heliocentric view of the universe. Newton discerned vestiges of this ancient rite in most human societies: thus he viewed such diverse structures as Stonehenge and the Jewish temple as examples of prytanea and he drew on accounts of contemporary India and the East Indies to provide instances of the survival of the cult of the vestal flame.[60] But with the growth of idolatry the true significance of this rite was lost and so, too, was the true natural philosophy. Gradually the sense of cosmic order was lost as

men were led by degrees to pay a veneration to these sensible objects and began at length to worship them as the visible seals of divinity ... For tis agreed that Idolatry began in ye worship of ye heavenly bodies & elements.[61]

Among the Egyptians — the major instigators of such idolatry — the vestal fire came to be regarded as an image of a fire in the centre of the earth rather than the sun, a corruption which was accompanied by the change from a heliocentric to a geocentric cosmology.[62]

Thus the re-establishemnt of the true heliocentric natural philosophy and the accompanying stripping away of idolatrous animistic conceptions of the way in which the universe worked was, for Newton, both a scientific and a religious mission. His work in natural philosophy and his theologico-historical investigations all shared the same goal: the restoration of that true understanding of God's relations with humanity and the natural order which had been obscured by the growth of idolatry after the time of Noah and his sons. Like Thomas Burnet,[63] Newton's disciple, William Whiston (and, very probably, the considerably more reticent Newton himself) regarded the recovery of the true natural philosophy as something of major religious, as well as scientific significance — indeed as ushering in the millenium. Thus Whiston viewed the publication of the *Principia*

as an eminent Prelude and Preparation to those happy Times of the Restitution of all things which God has spoken of by the Mouth of all his Holy Prophets since the world began, Acts, iii, 21.[64]

As Schaffer has shown,[65] one specific instance of Newton's belief that false conceptions of theology and natural philosophy were intertwined was his treatment of the problem of the movement of the comets. The view that comets operated only in the sublunar sphere — which, in Newton's view, had originated with Eudoxus (c. 427—c. 347 BC), who had studied under Plato and travelled in Egypt — was a corollary of the growth of idolatry. For the assumption that there were solid heavenly spheres which were supposed to prevent the movement of comets in the superlunary sphere derived from the belief that each heavenly body was associated with a god or spirit who moved the celestial sphere. By restoring the true understanding of the comets, then, Newton was attacking not only false astronomy but also what was, for Newton, the gravest of sins: idolatry. By re-establishing an understanding of the movement of the comets on true principles of natural philosophy and theology Newton also helped to contribute to the same goal as had prompted the *Discourse Concerning Prodigies* of his Cambridge contemporary, John Spencer: the undermining of the popular divination and superstition associated with comets by showing that the movements of comets were predictable and that their capacity to influence human affairs was limited by the fact that they belonged to the distant realm of the superlunary sphere. Newton also wished to show that the comets, and the heavenly bodies more generally, were merely inanimate servants of an all-wise and all-powerful Providence rather than, as idolatrous ancient astronomers had argued, the outward manifestations of a pagan pantheon.

Though it is highly probable that Newton knew and approved of Spencer's work on prodigies Newton does not appear to have made any specific reference to it. However, Newton's debt to Spencer's historical work — particularly his emphasis on the importance of Egyptian civilization for understanding the development of Judaism — is manifest. Newton plainly discussed issues related to the interpretation of the Old Testament with Spencer since, in a letter to Locke of 13 December 1691, he (Newton) remarked that Spencer had perused Le Clerc's controversial new Latin edition 'and likes ye design very well'.[66] After Spencer's death in 1693 Newton referred to him, in a letter to Otto Mencke, a professor at Leipzig, of 30 May as 'our friend' and also talked in some detail about the plans for publication of a second edition of Spencer's *De Legibus*.[67] Newton's library also included a well-thumbed copy of his work.[68] Most significantly, it was Spencer whom Newton cited in his *Theologiae Gentilis* to support the view that the religion of Moses was largely derived from that of the Egyptians 'for', as Newton wrote

Dr. Spencer has shown yt Moses retained all ye religion of ye Egyptians concerning ye worship of ye true God, & rejected only what belonged to ye worship of their false Gods . . . & that ye Mosaical religion concerning ye true God contains little else besides what was then in use amongst the Egyptians.[69]

Newton then used Spencer's painstaking scholarship in support of his basic hypothesis: that the true religion was that of Noah and his sons and that Judaism (and even Christianity) represented partial restorations of this pristine religion rather than wholly new religious dispensations. The Egyptians, he argued, had preserved an adultered form of the Noachian religion which they derived from Ham, one of the sons of Noah who had settled in the area.[70] Having invoked Spencer's evidence Newton continued the above passage by arguing that it followed that

it's certain that ye old religion of the Egyptians was ye true [Noachian] religion tho corrupted before the age of Moses by the mixture of fals Gods with that of ye true one.

Newton then concluded by vigorously reasserting the basic theme of the whole work — that it followed that the religion 'wch Moses taught ye Jews was no other than ye religion of Noah purged from the corruptions of ye nations'.[71]

Egypt, then, assumes a central importance in Newton's conception of world history for it had preserved the Noachian religion — albeit in an adulterated form — and transmitted it to the Jews who then went some way towards restoring it to its original form. Like Spencer's or Burnet's it is a conception of history which greatly restricts the extent of direct providential intervention in human affairs. God, in Newton's account,

'THE WISDOM OF EGYPTIANS' AND SECULARISATION

largely relies on his ordinary providence to achieve his purposes in history just as Newton also portrays Providence as chiefly working through secondary causes in nature. For both Newton and Spencer the religion of Moses was not, as a literal interpretation of the Old Testament would depict it, solely the result of direct divine revelation but rather was in large measure the product of the cultural assimilation by the Jews of certain aspects of Egyptian ritual practices. True, Judaism purged itself of many idolatrous Egyptian customs but this was a slow and by no means complete transformation — a further example of the way in which Providence was content to work through the natural processes of human history.

Elsewhere Newton, like Spencer, also suggested that aspects of Christianity developed out of Judaism by a similar process of historical evolution. Thus Newton emphasised the gradual way in which such Jewish practices as the appointment of a president of the synagogue were transformed into the Christian office of bishops.[72] The relationship of Christianity to Judaism, then, was to Newton akin to the relation between the religion of Moses and that of the Egyptians. As David Gregory noted after studying Newton's 'tract on the origin of nations, the Theologiae Gentilis' at Cambridge in 1694 Newton was of the view that 'Moses began a reformation but retained the indifferent elements of the Egyptians . . . Christ reformed the religion of Moses'.[73]

For Newton, then, the Egyptians played almost as central a role in the maintenance of the true Noachian religion as the Jews. This same tendency to emphasise the historical importance of the Egyptians in the transmission of religion and culture rather than, as had traditionally been the case, the Jews, can also be seen in other aspects of Newton's work. Where many sixteenth and seventeenth century scholars (including Theophilus Gale) argued that all languages derived from Hebrew, the original Adamic tongue,[74] Newton argued that the other languages of the Middle East derived from Egyptian rather than Hebrew.[75] Again Newton departs from the common apologetic device of tracing Greek learning back to the Jews either directly or via the Egyptians when he asserted in a draft introduction of Book III of the *Principia* that it was as a result of the astronomical observations of the Egyptians that the true, heliocentric natural philosophy

was spread abroad among other nations; for from them it was, and the nations about them, that the *Greeks*, . . . derived their first, as well as soundest, notions of philosophy.[76]

Where Newton's Cambridge contemporaries, the Platonists, Ralph Cudworth and Henry More, attempted to trace the atomic philosophy back to Moses Newton followed them only as far back as the Egyptians and the Phoenicians and, despite his close study of Cudworth's work, Newton was not persuaded by the Cambridge Platonist's identification of Moschus the Phoenician — the alleged founder of atomism — with Moses.[77] He did, however, draw heavily on the material which Cudworth assembled to demonstrate that the Egyptians 'excel[led] almost all other nations in reputations for wisdom, antiquity, and almost every other advantage'.[78]

But such admissions of Egypt's historical importance were made rather grudgingly by Newton for, in his historical *schema*, Egypt was not only central in the transmission of some elements of the true religion and the true natural philosophy but also central in the propagation of both religious and philosophical error. For Newton agreed abundantly with Cudworth's claim that

> there is no question to be made, but that the Greeks and Europeans generally derived their polytheism and idolatry from the Egyptians.[79]

As David Gregory noted after reading the *Theologiae Gentilis* in 1694 it was Newton's view that

> it was the Egyptians who most of all debased religion with superstition and from them it spread to the other peoples.[80]

In his other historical manuscripts Newton attributed the growth of such idolatry to the overweening ambitions of the Egyptian kings who sought to overawe their subjects by deifying their predecessors. It was this that led to what virtually takes the place of the Fall of Adam in Newton's theologico-historical system: the corruption of the true religion and, with it, the true natural philosophy of Noah. 'Here then', he writes of the actions of the Egyptian kings,

> we have the true original of ye corruption of the religion of Noah and the true cause of its spreading so early and so generally. For this policy of ye kings of Egypt soon took with ye kings of other nations.[81]

This corruption was compounded by the Egyptian priests who pandered to the ambitions of the kings and consolidated their own position by taking further this process of royal deification since they made reigns of those Gods very long & very ancient and separated them from the reign of the deified man.[82]

This idolatry also infected their study of astronomy for as idolatry became more and more established in Egypt as a result of royal and priestly ambitions so, too, priests began to attribute the movement of the heavens to the influences of the gods. As Newton wrote in his *Theologiae Gentilis*:

And by means of these fictions the souls of ye dead grew into veneration with ye stars & ... were taken for ye Gods, which governed this world.[83]

In another draft of the same work Newton caustically remarked

And so Astrology and gentile Theology were introduced by astute priests to promote the study of the Stars and to increase the power of the priesthood.[84]

Eventually, the result was the decline of the true, heliocentric philosophy and its replacement by the geocentric system which was subsequently codified by the Egyptian, Ptolemy.[85]

However, the significance which Newton attributes to the Egyptians for ill as well as good only serves to emphasise further their centrality in his conception of world history. The fall of the Egyptians into idolatry and their departure from the true natural philosophy provides Newton with virtually a partly secularised alternative to the biblical myth of the expulsion of Adam and Eve from Eden. For in Newton's account the original Fall has little significance — true religion, morality and philosophy could be found after Adam, and even after the Flood, in the age of Noah and his sons. The real fall from grace comes with the corruption of the Noachian religion as a result of the growth of idolatry — something for which Newton held the Egyptians to be chiefly responsible. In Newton's theologico-historical system, then, the Egyptians assume something of the central significance traditionally accorded to the Jews, except that in Newton's account the positive and negative features of Egypt's impact on the ancient world are explained largely in naturalistic terms. Thus the vital role that Newton attributes to the Egyptians in the transmission of the true Noachian religion to the Jews (albeit in an adulterated form) to some extent takes the place of a direct divine revelation to Moses while the original sin of idolatry is largely explained in terms of kingly and priestly ambitions rather than by recourse to the role of Satan.

However, the Egyptians loom so large in Newton's account almost against his will, for it was the aim of much of his historical writing to deflate the claims of the Egyptians and, conversely, to magnify the historical importance of the Jews. Indeed, Manuel sees most of Newton's chronological writings as a reaffirmation of the traditional historical *schema* in which the Jews were regarded as the true source of the civilization of the ancient world.[86] Where possible Newton did attempt to demonstrate the historical primacy of the Jews by undermining the Egyptian chronology as the work of ambitious priests working in league with kings wishing to instil obedience in their subjects by exaggerating the antiquity of their dynasty.[87] In his *Chronology of Ancient Kingdoms Amended* Newton devoted particular attention to demonstrating that the reign of Solomon was an historical watershed marking the beginnings of the great empires — a development which he argued occurred first in Israel.[88] Since this and a short chronological abstract based on it were the only historical works of Newton to be published he naturally came to be regarded as an advocate of the traditional claims for Jewish as against Egyptian priority. This the more so since in the *Chronology* Newton set out to argue that many of the attributes of civilization developed in Israel not long before the reigns of Solomon or his father, David, with the clear implication that other nations derived such skills from the Jews. Thus in discussing the age of mankind he wrote that:

And the original of letters, agriculture, navigation, music, arts and sciences, metals, smiths, and carpenters, towns and houses was not older in Europe than the days of Eli, Samuel and David; and before those days the earth was so thinly peopled, and so overgrown with woods, that mankind could not be much older than is represented in Scripture.[89]

But Newton's ambivalence in the last sentence of this passage is revealing. Where possible his historical writings are intended to bolster the biblical chronology but, in history, as in natural philosophy, Newton did not regard Scripture as providing an account which the true believer was obliged to follow without question or interpretation. And, as Westfall and Manuel point out,[90] the approach of Newton and other would-be reconcilers of biblical and classical chronologies, with their elaborate calculations designed to intermesh the biblical histories with data derived from the records of Gentiles and (in the case of Newton) astronomical phenomena, tended to undermine the privileged position of Scripture. Thus, in Newton's hands, the Bible frequently became but one ancient source among others to be used in a similar manner to the way one used the records of the Egyptians or the Greeks. Fabulous events in the Bible, as in the Gentile histories, were to be regarded — as Newton regarded the episode of the sun standing still at the battle of Jericho — as 'poetic expressions' which needed to be translated into secular historical terms. True, Newton's whole aim was to establish the ultimate authenticity of the biblical history but in doing so he drained it of much of its theological content by reducing it to the same level as other historical records. Thus in his *Chronology of the Ancient Kingdoms Amended* there is no hint of the biblical drama of salvation as Newton incorporates the scriptural history into a lifeless succession of kings and battles. Though Newton's purely chronological scripturally-based writings may have sought to assert the primacy of the Jews where, as in his *Theologiae Gentilis*, he moved beyond narrative to provide some sort of interpretative framework to explain the religious history of humankind the Eygptians — almost against Newton's wishes — are accorded a degree of historical significance which almost rivals that of the Jews.

Newton's ambivalent attitude to the Egyptians — like his somewhat cautious use of Scripture as an historical record — indicates the extent to which he, like Spencer, Burnet and other scholars of the age, reflects what Hazard[91] calls the 'crisis of the European conscience' of the late seventeenth century: the transition from a civilization based on an unquestioning acceptance of Christianity to one in which government, and social institutions more generally, were increasingly justified in more secular terms. This, in turn, reflects a growing wariness about linking such institutions too closely to a specifically religious conception of the world in order to avoid the confessional strife unleashed by the Reformation. Newton's almost instinctive hostility to the Egyptians derived from a long-established form of Christian apologetic — the argument that Gentile learning (particularly that of the Egyptians and the Greeks) ultimately derived from Jewish sources. As Bernal points out,[92] this long-engrained tendency to downplay the importance of Egyptian culture was further strengthened for Newton and many of his contemporaries by the fact that anti-Christian writers ranging from Giordano Bruno in the sixteenth century to the English Deists of the early eighteenth century had looked to ancient Egypt to provide their beliefs with the seal of antiquity.

On the other hand, however, Newton's natural tendency to extend to the history of the Judaeo-Christian tradition the emphasis on the role of secondary agents which had produced such explanatory success in his natural philosophy brought with it, however reluctantly, an ineluctable tendency to accord the Egyptians great significance in the history of religion and civilization. Moreover, as a close student of the ancient chronicles Newton found it difficult to downplay the role of the Egyptians to the extent demanded by the traditional interpretation which accorded the Jews almost total historical primacy. In any case, as Spencer, Burnet and others had pointed out, such a view was not clearly stated in Scripture and, indeed, could be challenged on the basis of Scripture itself — for did not the Bible speak of Moses drinking deeply at the founts of Egyptian wisdom? Newton's ambivalent attitude towards the Egyptians was even reflected in his discussion of points of historical detail: sometimes he suggests that writing derived from the Jews but elsewhere he attributes it to the Egyptians;[93] though in his *Chronology of the Ancient Kingdoms Amended* he makes much of the significance of the fact that Solomon's reign preceded that of the great dynasties of Egypt, in the *Theologiae Gentilis* he appears to question this by describing Egypt as 'ye oldest of kingdoms'.[94]

Newton's tendency — however reluctantly — to accord the Egyptians greater significance in world history indicated an attitude to the role of Providence which was also reflected in his natural philosophy. Thus his greater attention to the role of the Egyptians, rather than the Jews, in the transmission of the true religion and of the arts of civilization betokened a greater emphasis on the extent to which Providence worked in history through secondary causes rather than by means of direct intervention. Similarly, in his natural philosophy he and, *a fortiori*, his clerical popularisers such as Bentley, Clarke and Whiston, argued that his natural philosophy demonstrated the way in which Providence used such secondary agents as gravitation or the movement of the comets to achieve that harmony and order which, to Newton and his disciples, was the hallmark of the work of the Creator.

However, just as the Deists discarded the role of a continuing and active Providence in Newton's system of natural philosophy leaving God to act merely as the First Cause — the distant and remote Watchmaker God — so, too, some of the Deists argued for an understanding of world history devoid of providential direction. The Deists also sought to undermine the significance of Scripture by relegating the role of the Jews in the ancient world to one of almost total insignificance and, conversely, by magnifying the role of the Egyptians. The most influential statement of this overtly anti-Judaeo-Christian historical interpretation was to be Voltaire's *Philosophy of History* written as an explicit attack on the theocentric conception of world history embodied in Bishop Bossuet's influential *Discours sur L'Histoire Universelle* (1681)[95] — a work which took as its guiding theme for the history of the ancient world the biblical account of the history of the Jews as 'a palpable account of his [God's] external providence'.[96] By contrast, Voltaire implicitly rejects both the antiquity and the historical reliability of the Old Testament by writing of the Chinese chronicles that 'It is incontestable that [they are] the most ancient annals of the world' and praising them for providing 'an uninterrupted succession, circumstantial, complete, judicious, without any mixture of the marvellous'. With characteristic impishness he dismisses Genesis' account of the peopling of the world after the Flood with the aside that

I shall leave to men more learned than myself, the trouble of proving that the three children of Noah, who were the only inhabitants of the globe, divided the whole of it amongst them.

In a chapter entitled 'Whether the Jews taught other Nations, or whether they were taught by them' Voltaire, predictably, sets out to show the extent to which the Jews derived their culture from the Gentiles and, particularly, the Egyptians.[97]

But just as Voltaire used Newton's natural philosophy — which Newton and most of his English disciples considered as a natural aid to Christian apologetics — as a Deistic weapon against the Church so, too, Voltaire's overtly Deistic system of world history ultimately largely derived from the work of English Christian scholars such as Spencer. Another important late seventeenth century scholar who arrived at similar conclusions to those of Spencer about the relationship between ancient Egyptian and Jewish culture was Sir John Marsham, author of *Chronicus Canon Aegyptiacus, Ebraicus, & Graecus* (1672). Like so much historical scholarship of the period this work aimed to achieve a peaceful reconciliation between the biblical, classical and Egyptian chronologies thus undercutting what Bishop Stillingfleet called one of

the most popular pretences of the Atheists of our Age . . . the irreconcilableness of the account of Times in Scripture, with that of the learned and ancient Heathen Nations.[98]

In the preface to this work Marsham, at pains to emphasise his orthodoxy, argued that his view of the relations between the Jews and the Egyptians was in accord with that of many of the Church Fathers and with the *Dissertatio de Urim et Thummim* (1669) — which he praises for its exhaustive scholarship[99] — Spencer's early work (which was later incorporated into the *De Legibus*) in which he demonstrated the extent of Egyptian influence on Jewish divinatory practices. Marsham's work did much to provide scholars with a coherent survey of Egyptian history and was used extensively: Spencer drew on it in his *De Legibus* and Newton often cites his work — it was Marsham, for example, who was Newton's authority for the (false) identification of the purportedly early pharoah, Sesostris, with the figure of Sesac who is mentioned in the Bible as dating from after the death of Solomon.[100]

Marsham's work — along with other Christian apologists like Spencer and Burnet — was, however, put to very different uses by a number of the English Deists as part of their assault on the Old Testament. Along with Burnet's *Archaeologia Philosophica* — the influence of which he enthusiastically acknowledges — it was very likely Marsham and Spencer on whom Toland was drawing when, in his *Letters to Serena* (1704), he attacked the view that 'the Jews and a world of Christians pretend that the Egyptians had all their learning from Abraham'. By contrast, Toland argued that 'the Jews were of all Eastern People the most illiterate' and that it was the Egyptians who were 'the Fountains of Learning to all the East'.[101] In his *Christianity as Old as Creation* (1730) Toland's fellow Deist, Matthew Tindal, cited 'Marsham, and Others' as his authority for asserting that the Jews derived the ritual of circumcision from the Egyptians. For Tindal this was one illustration of the thesis embodied in his title: that neither Moses nor Christ instituted a new religious order since the true religion should have been evident from the beginnings of humankind to anyone of good will and sound reason — in short, that 'True Religion, in all Places and Times, must ever be the same; Eternal, Universal and Unalterable'.[102] Such was also the thesis of Newton's *Theologiae Gentilis* (though there is nothing to suggest any mutual influence) with the difference that, in Newton's work, Moses and Christ are still accorded a significant (albeit much reduced) role in the restoration of the true, original religion which, for Newton, was that of Noah and his sons. Tindal's reference to 'Marsham, and Others' probably embraced Spencer since, although the author of *De Legibus Hebraeorum* does not definitely state that the Jews derived the practice of circumcision from the Egyptians, he assembled a great deal of evidence which suggested that this was highly probable.[103]

Tindal's work prompted a sermon entitled 'The Wisdom of the Ancients Borrowed from Divine Revelation, or, Christianity Vindicated Against Infidelity' (1731) from that tireless defender of orthodoxy, Daniel Waterland. As the title suggests this was a restatement — though couched in very cautious terms — of the ancient argument that the Gentiles, whether Egyptian or Greek, had derived their learning from the Jews. Though Waterland is gentle in his treatment of two such venerable scholars as Marsham and Spencer he makes it clear that he regards their work as having provided ammunition for Deistic attacks on the Old Testament such as that of Tindal. While the ancient argument for the priority of Jewish learning may have been overstated, he writes,

Two very considerable writers, Sir John Marsham and Dr. Spencer, appear to have slighted it too much. They have not only called in question the prevailing opinion of the ancient Apologists, but they have run directly counter to it; pretending that the Pagans did not borrow from the Jews, but that the Jews rather copied after the Egyptians or other Pagans, in such instances as both agree in: a strange way of turning the tables, confounding history, and inverting the real order of things.[104]

Waterland's work prompted, in turn, a series of pamphlets from his Cambridge colleague, Conyers Middleton, whose religious beliefs appear to have hovered uncertainly between a form of enlightened Christianity and a mild Deism — Gibbon wrote of him that he 'rose to the highest pitch of scepticism, in any wise consistent with religion'.[105] In the first of these pamphlets, *A Letter to Dr. Waterland* ... (1731), Middleton vigorously asserted that the defence of Christianity was ill-served by Waterland's form of apologetics for anyone familiar with the ancient world knew that '*Aegypt was a great and powerful nation*' when the Jews were '*an obscure contemptible people,* famed for no kind of literature; scarce known to the polite world till the Roman Empire dispersed them'. Moreover, the Greek historians clearly favoured the view that it was the Jews who borrowed such ritual practices as circumcision from the Egyptians rather than, as Waterland maintained, the other way around

and twas the authority of these that induced the *learned Marsham*, and the no less *learned Spencer* too, to favour the opinion of your adversary [Tindal].[106]

In a subsequent pamphlet directed at one of Waterland's advocates Middleton reasserted his view that the fact that the Jews derived many of their rituals from the Egyptians had been so clearly demonstrated by the learned Spencer that

'no Man, unless supremely credulous', as he says, 'can believe it have been the Egyptians'. And indeed both he and Marsham derive in a manner the whole ritual Law from this very source of Egypt.[107]

Spencer's and Marsham's works also provided much of the scholarly foundation for Warburton's ill-digested compilation of learning, *The Divine Legation of Moses Demonstrated* (1741). This was written, as Rossi[108] points out, in an attempt to turn the attacks of the Deists (and especially John Toland) on Christianity back on themselves by using their claims about the relationship between the Jews and the Egyptians in the defence of orthodoxy. In the *Divine Legation* Warburton advances the paradoxical argument that the Jews plainly received a Divine Revelation which instilled in them a strong belief in an active Providence who rewarded good and punished evil since — unlike the Gentiles and, in particular, the Egyptians — they were able to maintain social and political order without the sanction of a belief in an after-life. This was particularly remarkable since in other respects, writes Warburton (echoing Spencer and Marsham),

the Jews were extremely fond of Egyptian manners and did frequently fall into Egyptian superstition, and that many of the Laws given to them by the Ministry of Moses were instituted, partly in compliance to their prejudices and partly in opposition to those superstitions.[109]

Elsewhere Warburton makes explicit his debt to Spencer by praising 'the learned SPENCER' as one who

hath fully exhausted this subject in his excellent work ... and thereby done great service to divine revelation: for the RITUAL LAW, when thus explained, is seen to be an institution of the most beautiful and sublime contrivance.

Ultimately, he argued, Spencer's thesis could be reduced to the proposition that '*Moses gave the ritual law to the Jews because of the hardness of their hearts*; the very hypothesis of Jesus Christ'. Like Spencer, then, Warburton saw the influence of the Egyptians on the Israelites as having been providentially ordained though it was an example of Providence's use of secondary agents rather than of direct divine intervention. Similarly, he saw the preservation of Egypt 'from total destruction', despite its role in promoting idolatry, as part of

Providence's design to ensure the diffusion 'of civil life and polished manners, which were to derive their source from thence'.[110]

Far from arguing in the traditional manner that the Egyptians derived their learning from the Jews it was an important part of Warburton's highly ingenious form of historical apologetics that Egyptian civilization was older than that of the Jews and that the Jews drew heavily upon it. For, in Warburton's opinion, this made it all the more remarkable that the Jews did not adopt the Egyptian belief in an afterlife and, consequently, strongly suggested that they were the beneficiaries of a special revelation which ensured a belief in an active Providence and thus enabled them to maintain good behaviour and civil order without such an essential sanction as fear of retribution in a life to come. Warburton therefore devoted much space to confuting Newton as one of the chief exponents (at least in his published works) of the view that Jewish civilization was older than that of the Egyptians. Such a position, Warburton acknowledged, put himself, as a Christian apologist, in some rather odd company since 'the present turn, indeed, of free-thinking is to extol the high antiquity of Egypt, as an advantage to their cause'. Nonetheless, Warburton was confident that as his contemporaries came to appreciate the force of his strikingly original argument and the orthodox emphasised the debt of Israel to Egypt so, too 'we shall see the contrary notion, of the low antiquity of Egypt become the fashionable doctrine' among non-believers.[111]

The aim of Warburton's work, like that of Spencer's, was, then, to illustrate the way in which God's general Providence could use the natural processes of history — such as the influence of the Egyptians on the Jews — to achieve his ends. However, the basic apologetic intent of the work is also evident in the way in which Warburton emphasises the extent to which such historical events were not, as the Deists claimed, essentially random but rather were firmly under the direction of Providence, even though it might be acting through secondary causes. Similarly, in his depiction of the relationship between God and the natural world Warburton was at pains to emphasise that, although God generally worked through what he called 'the stated laws of Physics', nonetheless these were also part of a divinely-planned, providential order. Thus in his sermon, 'Natural and Civil Events the Instruments of God's Moral Government' — preached in 1755 on the Fast Day called in response to the Lisbon earthquake — Warburton argued that it was a 'disrespectful notion of divine Wisdom' to maintain that God could only manifest his system of rewards and punishments through miracles rather than through the natural order since

> God, by the admirable direction of his general providence, so adjusts the circumstances of the natural and moral system, as to make the events in the former to serve for the regulation of the latter.[112]

Although Warburton's understanding of God's relation to history and nature emphasised the extent to which God worked through his general or ordinary Providence he was also anxious to leave room for God's exercise of his extraordinary Providence — in short, for miracles. For in the *Divine Legation* the fact that the Jews maintained civil order without a belief in an after-life is, to Warburton, a clear indication of the fact that they were exposed to overt manifestations of God's extraordinary Providence which made them obey the moral law without any other sanction. In another of his works revealingly entitled, *Julian: or A Discourse concerning the Earthquake and Firey Eruptions, which defeated that Emperor's Attempt to rebuild the Temple at Jerusalem. In which the Reality of a Divine Interposition is shewn* ... (1750), Warburton set out to document a miracle of the fourth century AD 'worked by the immediate hand of God, and not through the agency of his servants'[113] to counter the claims of Middleton that any purported miracles after the apostolic period were likely to be spurious. Consistent with such an understanding of history, in which the miraculous still played a significant role, Warburton was reluctant to treat many of the historical passages of the Bible as being allegorical in the way in which passages dealing with natural philosophy had been reinterpreted for, as he wrote in *The Divine Legation* in relation to Newton's historical work,

> though the end of the sacred history was certainly not to instruct us in astronomy yet it was, without question, written to inform us of the various fortunes of the people of God; with whom, the history of Egypt was closely connected.[114]

Nonetheless, though Warburton was wary of allowing the Deists an opening to eliminate the role of direct providential intervention in history altogether, the overall drift of his historical writing was to give greater emphasis to God's ordinary rather than his extraordinary providence. Appropriately, Warburton's youthful work (his second) — *A Critical and Philosophical Enquiry into the Causes of Prodigies and Miracles as Related by Historians* (1727) — was concerned to cleanse profane history of many supposed instances of supernatural intervention while providing rules for distinguishing true miracles from the false.[115]

But, ironically, just as Warburton had attempted to turn the tables on his Deistic opponents by using their claims about the greater antiquity of Egyptian culture as against that of the Jews in defence of orthodoxy so, too, Voltaire attempted to use Warburton's apologetic labours as part of his Deistic assault on Christianity. For it was largely from Warburton and other English populisers of the scholarly labours of Marsham and Spencer that Voltaire drew his material on the relations between the Egyptians and the Jews which, in his *Philosophy of History*, formed part of his Deistic assault on the traditional Judaeo-Christian conception of history which had received its last great statement in Bossuet's *Discours sur L'Histoire Universelle*. As Torrey and Brumfitt[116] have shown, Voltaire had closely studied Warburton's *Divine Legation* and Middleton's pamphlets against Waterland before writing the *Philosophy of History* in 1765. He was also greatly influenced by Tindal's critical approach to the Bible in *Christianity as Old as Creation*. By such an indirect route, then, did the scholarship of Spencer and Marsham, which had been intended to draw together into a harmonious whole the biblical and classical historical sources, come to serve the purposes of the eighteenth century's most influential opponent of revealed religion. By a similar irony the natural philosophy of Newton which its author and his early English disciples saw as a confirmation of the existence of a Christian Providence was also to be used by Voltaire as another weapon in the crusade to *'écrasez l'infâme'*.

But Voltaire's work represented the response of a French intellectual class which was profoundly alienated from its society and keen to press into service any potential means of attack on an ossified Church and State. In eighteenth century England, however, as the anti-clericalism which lay at the root of the Deists' work abated once the Whigs took firm control of the Church so, too, did the tendency to use history (or natural philosophy) in order to undercut the power of priestcraft also begin to wane. The belief that the normal processes of nature were controlled by a directing Providence became so closely associated with the dominant Newtonian natural philosophy that science and Christian apologetics came to be seen as natural allies.[117] As such a view gained acceptance so, too, did the principle that the scriptural account of natural processes frequently needed to be treated as being couched in metaphorical language appropriate to the needs of a pre-scientific people — for such a view had been advocated by Newton himself and was reflected in his natural philosophy. The increasing sway of Newtonian natural philosophy thus helped to strengthen two principles which had considerable importance for a more secular understanding of history: an emphasis on God's ordinary, as against his extraordinary Providence, and a willingness to interpret Scripture metaphorically. As such principles gained ground such ancient problems as the relationship between the Jews and the Egyptians came less and less to be seen as central issues in the defence of revealed religion. For once it was accepted that Providence worked in history, as in natural philosophy, primarily through secondary agents — such as the influence of the Egyptians on the Jews — and that the creation of the Mosaic law, like the Mosaic cosmology, should not be interpreted literally as the immediate result of divine intervention, than such issues lost much of their urgency. Lord Monboddo, for example, prompted little controversy (at least, of a theological kind) when he argued in the late eighteenth century in his *Of the Origin and Progress of Language* (1773—92) that Egypt was the native country of all

Arts, Science, and Philosophy, and ... from thence they have been derived to all the Nations[118]

— a claim that would once have been regarded as undermining the central historical importance of the Jews.

Furthermore, as it came to be appreciated the extent to which the religion of the ancient Israelites was influenced by their Gentile neighbours and, in particular, by the Egyptians, so too the view was strengthened that true religion, like other areas of human experience, had developed historically and was, in large measure, the daughter of time. William Robertson, who saw no conflict in combining his role as Moderator-General of the Church of Scotland with his pioneering work in establishing history as a secular discipline, discoursed on this theme in his sermon entitled 'The Situation of the World at the Time of Christ's Appearance, and its Connexion with the Success of his Religion, Considered' delivered in 1775. In it he emphasised the extent to which Providence 'conducteth all his operations by general laws' arguing that 'The Almighty seldom effects by supernatural means anything which could have been accomplished by such as are natural'.[119] For

Robertson an important illustration of this general theme was the extent to which God had relied on natural historical processes in the development and promulgation of Christianity. 'The light of revelation', he writes, 'was not poured in upon mankind all at once, and with its full splendour'. Rather the development of true religion was conditioned by such natural phenomena as the rise and fall of empires for

In proportion as the situation of the world made it necessary the Almighty was pleased further to open and unfold his scheme. And men came by degrees to understand this progressive plan of Providence.[120]

In Robertson's account the outstanding example of the way in which Christianity had benefited from such natural causes was the fact that it had largely been disseminated through the agency of the Roman Empire. Conversely natural causes could also play a role in retarding true religion: Robertson partly attributed what he saw as the persistence of pagan residues in ancient Judaism to the effects of climate along with the influence of the Israelites' Gentile neighbours.[121] As such an historically-based understanding of the development of Christianity gained ground so, too, the literal account of world history outlined in the Bible came to be seen as being itself a reflection of particular historical circumstances which required the skills of an historian to explain and to illuminate.

Consequently, although historical issues still remained of considerable importance for an historically-based religion such as Christianity it increasingly came to be acknowledged that history — including history of the ancient world which traversed similar ground to that outlined in the Bible — like natural philosophy, was a field which ought to be distinguished from theology. Of course it was assumed by believers that ultimately the fruits of secular history would be in accord with Scripture, rightly understood, and that secular history, like natural history, revealed the natural mechanisms by which Providence achieved its ends. Nonetheless, the Bible itself was less and less assumed to be the ultimate foundation on which histories of humankind should be based. Just as it came to be accepted that the Bible had not been intended to provide a system of natural philosophy so, too, it became widely, though by no means universally, acknowledged that the biblical narrative was not meant to supply the sort of historical framework which the age of the Enlightenment came to demand — one which provided an explanation of the origins of human society and institutions in terms of natural causes and which, as far as possible, used the methods so successfully employed in the natural sciences. Increasingly, then, biblical history became the pattern which was invoked to give meaning to an individual's religious experience rather than to provide a framework for explaining the development of the ancient world.[122] Moreover, as knowledge expanded, the traditional impulse to unite all fields of learning under the aegis of theology began to weaken, thus strengthening the autonomy of history as a discipline with different methods and goals to that of biblical theology. Thus that movement of ideas to which we give the name the Scientific Revolution involved more than a change in humanity's attitude to the natural world for it also helped to tranform our attitude to our past and in so doing helped to bring to birth the discipline of history in its modern secularised form.

NOTES

Note Added in Proof.
J. E. Force and R. H. Popkin (eds), *Essays on the Context, Nature and Influence of Sir Isaac Newton's Theology*, Dordrecht (1990), which includes a number of valuable articles relevant to this essay, appeared too late to be used in its preparation. For some comments see my review of this work in *Physis* (forthcoming).

[1] Funkenstein, A. *Theology and the Scientific Imagination*, Princeton (1986).
[2] For a useful review of the literature on this subject see Preston, J., 'Was There an Historical Revolution', *Journal of the History of Ideas* 38, 353—64 (1977).
[3] Smith Fussner, F., *The Historical Revolution: English Historical Writing and Thought 1580—1640*, London (1962), especially pp. 25, 297, 306—8 and Sypher, G. W., 'Similarities between the Scientific and the Historical Revolutions at the End of The Renaissance', *Journal of the History of Ideas* 26, 353—68 (1965).
[4] Manuel, F., *Isaac Newton Historian*, Cambridge, Mass. (1963), p. 29 and McGuire, J. E. and Rattansi, P. M., 'Newton and the "Pipes of Pan"' *Notes and Records of the Royal Society of London* 21 (1966), pp. 128—9. An early eighteenth-century sermon entitled 'The Wisdom of the Ancients Borrowed from Divine Revelation or, Christianity Vindicated against Infidelity' by the Anglican divine, Daniel Waterland, provides a useful survey of this form of historical interpretation complete with detailed references on this point in the works of Church Fathers such as Justin Martyr, Clement of Alexandria, Tertullian, Origen and Augustine. Van Mildert, W. (ed.), *The Works of the Reverend Daniel Waterland, D. D.* 2nd ed. 6 Vols, Oxford (1843), V, pp. 4—13.
[5] McGuire, J. E., 'Neoplatonism and Active Principles: Newton and the Corpus Hermeticum', in R. S. Westman and J. E. McGuire, *Hermeticism and the Scientific Revolution*, Berkeley (1977), p. 128.
[6] Manuel, *Newton Historian*, p. 101.
[7] Gale, T., *The Court of the Gentiles: or a Discourse Touching the Original of Human*

Literature, both Philologie and Philosophie, from the Scriptures and Jewish Church 2nd ed., Oxford (1672—82), Part I, p. [i], heading to Ch. X; Part II, pp. 39, 44.

[8] Wallis, J., *Three Sermons Concerning the Sacred Trinity*, London (1691), p. 99.

[9] Spencer, J., *De Legibus Hebraeorum Ritualibus et Earum Rationibus, Libri Quatuor* 2nd ed., Cambridge (1727), II, pp. 640—1. Translated in Woodward, J., 'Of the Wisdom of the Antient Egyptians; a Discourse concerning their Arts, their Sciences, and their Learning: their Laws, their Government, and their Religion. With occasional Reflections upon the State of Learning, among the Jews; and some other Nations' *Archaeologia* 4, p. 268 (1777).

[10] Spencer, *De Legibus*, II, p. 649. Translated Woodward, 'Wisdom of the Antient Egyptians', p. 269.

[11] Spencer, *De Legibus*, II, p. 1089.

[12] Literal translation from *ibid.* II, p. 741.

[13] Spencer's sermon attracted considerable attention to judge from the comments of that connoisseur of sermons, Samuel Pepys, who, on 1 June 1664, recorded reading 'Mr. Spenser's Book of Prodigys' and finding it 'most ingeniously writ, both for matter & style'. Pepys later notes that on 25 May 1666 he was out walking with a friend and 'discoursing & admiring of the learning of Dr. Spenser'. Latham, R. and Matthews, W. (eds), *The Diary of Samuel Pepys* 10 Vols, London (1970— 83), V, p. 165 and VII, p. 133. Spencer's sermon was republished in a second and considerably enlarged edition in 1665. In the same year he also published *A Discourse Concerning Vulgar Prophecies: Wherein the Vanity of Receiving Them as the Certain Indications of any Future Events is Discoursed; And Some Characters of Distinction Between True and Pretending Prophets are Laid Down* which, as the name suggests, was again directed at popular enthusiasm and its threat to political and religious order.

[14] Spencer, J., *A Discourse Concerning Prodigies: Wherein the Vanity of Presages by Them is Reprehended, and their True and Proper Ends Asserted and Vindicated*, Cambridge (1663), pp. [vi], [iii].

[15] *Ibid.*, [i].

[16] *Ibid.*, pp. 14—6.

[17] *Ibid.*, p. 43.

[18] *Ibid.* 2nd. ed., London (1665), p. 136.

[19] *Aegyptica . . . sive de Aegyptiacorum Sacrorum cum Hebraicis Collatione Libri Tres . . .*, Amsterdam (1683). Cited in translation in Warburton, W. *The Divine Legation of Moses Demonstrated* 2 Vols, London (1837), II, p. 167.

[20] Edwards, J. *Compleat History*, pp. 249, 252.

[21] Edwards, J. *Cometomantia*, London (1684), pp. 94, 98, 28.

[22] In 1682, Bayle published (in French) his *Letter to M. L. A. D. C., Doctor of the Sorbonne. Wherein it is Proved in the Light of Various Arguments Derived from Philosophy and Theology that Comets Are in No Sense Portents of Disaster . . . .* A sequel followed in 1683 with further supplements in 1694 and 1705. On the significance of Boyle's works see Hazard, P. *The European Mind 1680—1715*, (Cleveland, Ohio (1963), pp. 155—161.

[23] Edwards, J. *Cometomantia*, p. 131. Among the arguments which Edwards attacks is one developed by Spencer (though Edwards, understandably, does not explicitly cite his distinguished, senior Cambridge colleague on the point) — namely, Spencer's conten-

tion that comets could not be intended as a divine warning to a particular nation since comets were visible to many different countries simultaneously. *ibid.*, p. 158 and Spencer, *Prodigies* (1663), p. 17.

24 Woodward, 'Of the Wisdom of the Antient Egyptians', pp. 280, 225.
25 *Ibid.*, pp. 218, 228—9, 238, 281.
26 *Ibid.*, p. 271.
27 This work was published in two parts. The Latin first part appeared in 1681 (and was translated into English in 1684) and the Latin second part was published in 1689 and translated in 1690. Jacob, M. C. and Lockwood, W. A., 'Political Millenarianism and Burnet's *Sacred Theory*', *Science Studies* 2, p. 265 (1972).
28 Burnet, T. *The Theory of the Earth* 2nd ed., London (1691), p. 6.
29 *Ibid.*, pp. 289, 314—5.
30 Edwards, J. *A Demonstration of the Existence and Providence of God* . . . (1696), p. 257. Cited in Levine, J. M. *Dr. Woodward's Shield: History, Science and Satire in Augustan England*, Berkeley (1977), p. 265.
31 Woodward, J. *An Essay Toward a Natural History of the Earth*, London (1695), p. 165.
32 Cited in Porter, R. S. *The Making of Geology. Earth Science in Britain 1660—1815*, Cambridge (1977), p. 76.
33 Burnet, *Theory of the Earth*, p. 281.
34 Burnet, T. [*Archaeologiae Philosophicae sive*] *Doctrina Antiqua de Rerum Originibus: Or, An Inquiry into the Doctrine of the Philosophers of all Nations, Concerning the Original of the World* Translated by Mr. Mead and Mr. Foxton, London (1736), pp. 45, 56, 241, 244, 246. For an important discussion of Burnet's writings see Rossi, P. *The Dark Abyss of Time. The History of the Earth and the History of Nations from Hooke to Vico*, Chicago (1984), pp. 33—41, 66—74, 89—94.
35 Burnet, T. *An Answer to the Late Exceptions Made by Mr. Erasmus Warren Against the Theory of the Earth*, London (1690), p. 84.
36 Burnet, [*Archaeologiae Philosophicae*], p. 2.
37 *The Seventh and Eighth Chapters of Dr. Burnet's Archiologiae [sic] Philosophicae, together with his Appendix to the Same* . . . *Rendered into English, by Mr. H. B.* [Henry Brown] in Blount, C. *The Miscellaneous Works*, London (1695), pp. 29, 32—3, 39.
38 Cited in Redwood, J. *Reason, Ridicule and Religion: The Age of Enlightenment in England, 1660—1750*, London (1976), p. 119.
39 *The Seventh and Eighth Chapters of Dr. Burnet's Archiologiae Philosophicae*, p. 50.
40 Redwood, *Reasons, Ridicule and Religion*, p. 121.
41 *The Seventh and Eighth Chapters of Dr. Burnet's Archiologiae Philosophicae*, p. 54.
42 Cited in Kubrin, D. *Providence and the Mechanical Philosophy: The Creation and Dissolution of the World in Newtonian Thought*, PhD thesis, Cornell University (1968), pp. 145—6. The extent of clerical hostility to Burnet's work is indicated by the remark in 1693 of Humphrey Prideaux, canon of Norwich and author of an historical account of the period between the Old and the New Testaments, that the coffee-house atheists made much use of the *Archaeologiae Philosophicae* 'to confute ye account ye Scriptures give us of ye creation of ye world'. Thompson, E. M. (ed.), *Letters of Humphrey Prideaux, Sometime Dean of Norwich to John Ellis* . . . . Camden Society Publications, n.s. XV, London (1875), pp. 162—3.

43 Edleston, J. (ed.), *Correspondence of Sir Isaac Newton and Professor Cotes* ..., London (1850), p. xliv.
44 Turnbull, H. W. et al. (eds), *The Correspondence of Sir Isaac Newton* 7 Vols, Cambridge (1959—77), II, p. 331.
45 Burnet, *Theory of the Earth*, p. 315.
46 Turnbull, *Correspondence of Newton*, II, p. 334.
47 Harrison, J. *The Library of Isaac Newton*, Cambridge (1978), p. 112.
48 Westfall dates the beginnings of Newton's most important theologico-historical work, the *Theologiae Gentilis* from the mid 1680s although he returned to it in the early 1690s (at much the same time as Burnet's work appeared). Newton thereafter continued to revise the work until at least 1716. Westfall, R. S. 'Isaac Newton's *Theologiae Gentilis Origines Philosophicae* in W. W. Wagar, (ed.), *The Secular Mind. Transformations of Faith in Modern Europe*, New York (1982), pp. 15—34.
49 *Ibid.*
50 Westfall, R. S. *Never at Rest. A Biography of Isaac Newton*, Cambridge, (1983), p. 812.
51 Newton, I. *Opera Quae Exstant Omnia* ed. S. Horsley, 5 Vols, London (1785), V, pp. 140—1.
52 Cited in Westfall, 'Newton's *Theologiae Gentilis*', p. 30.
53 McLachlan, H. (ed.), *Sir Isaac Newton: Theological Manuscripts*, Liverpool (1950), p. 52.
54 Burnet, T. *The Faith and Duties of Christians* translated into English by Mr. Dennis, London [1728], p. 57.
55 Newton, I., *The Opticks*, Great Books, Chicago (1952), p. 542.
56 Kubrin, D. 'Newton and the Cyclical Cosmos: Providence and the Mechanical Philosophy', *Journal of the History of Ideas* 28, 325—46 (1967).
57 Newton, *Opticks*, p. 543.
58 McLachlan, *Newton: Theological Manuscripts*, p. 28.
59 Newton, *Opticks*, pp. 543—4.
60 Jewish National and University Library, Yahuda MS 41, fols 1—3.
61 *Ibid.* fol. 8.
62 Westfall, 'Newton's *Theologiae Gentilis*', p. 26.
63 See ref. 34 above.
64 Whiston, W. *Memoirs of the Life and Writings of Mr. William Whiston* ... 2 Vols., London (1749), I, p. 38. *Acts* 3:20—1 reads: 'And he shall send Jesus Christ ... whom the heavens must receive until the times of restitution of all things, which God hath spoken by the mouth of all his holy prophets since the world began'.
65 Schaffer, S. 'Newton's Comets and the Transformation of Astrology', in P. Curry (ed.), *Astrology, Science, and Society: Historical Essays*, Woodbridge (1987), pp. 219—43.
66 Turnbull, *Correspondence of Newton*, III, p. 185.
67 *Ibid.*, p. 292. The second edition did not, however, eventuate until 1727 when, interestingly enough, it was published with the assistance of a bequest from the latitudinarian Archbishop Tenison of Canterbury, a former fellow of Spencer's college — an indication that Spencer's work was by no means universally regarded with suspicion by his fellow clergy.

[68] Harrison, J. *Library of Isaac Newton*, p. 242.
[69] Yehuda MS 41, fol. 5.
[70] *Ibid.* fol. 10. Newton's Cambridge contemporary, Ralph Cudworth, argued that not only had Ham settled in Egypt but that the name of the Egyptian supreme god, Ammon (or Hammon), was derived from Ham. Cudworth, R. *True Intellectual System* 3 Vols, London (1845), I, p. 572, (first edition, 1678).
[71] Yehuda MS 41, fol. 5.
[72] Manuel, *Newton Historian*, p. 148.
[73] Turnbull, *Correspondence of Newton*, III, p. 338.
[74] On this point see Allen, D. C. 'Some Theories of the Growth and Origin of Language in Milton's Age', *Philogical Quarterly* **28**, 5–16 (1949), and Kottman, 'Fray Luis de Léon and the Universality of Hebrew: An Aspect of Sixteenth and Seventeenth Century Language Theory', *Journal of the History of Philosophy* **13**, 297–310 (1975).
[75] Westfall, 'Newton's *Theologiae Gentilis*', p. 23.
[76] Newton, I. *The Mathematical Principles of Natural Philosophy and his System of the World* translated by Andrew Motte. 2 Vols, Berkeley (1974), II, p. 549.
[77] Thus in his unpublished 'classical scholia' to the *Principia* Newton traces the atomic philosophy only as far back as Moschus the Phoenician from whom it passed to the Egyptians and thence to the Greeks. Casini, P. 'Newton: The Classical Scholia', *History of Science* **22**, p. 36 (1984). In his extensive notes 'Out of Cudworth' (now in the William Andrews Clark Memorial Library, University of California, Los Angeles) Newton copied much of Cudworth's material on Moschus but ignores his identification of Moschus with Moses. On this identification see Sailor, D. B., 'Moses and Atomism', *Journal of the History of Ideas* **25**, 3–16 (1964).
[78] Cudworth, *True Intellectual System*, III, p. 185. In one respect Newton went even further than Cudworth in praising the Egyptians for in the Clark MS he wrote of the Egyptians' natural philosophy (which he interpreted as embodying a mystical form of atomism) that 'Dr. Cudworth therefore is much mistaken when he represents this Philosophy as Atheistical' (lines 31–2). A recent study of this manuscript concludes that:

Newton was very much convinced of the priority and thus the importance of Egyptian learning in the ancient world, and this is reflected in almost an entire page of the references and quotes which he took from Cudworth.

Sailor, D. B. ' Newton's Debt to Cudworth', *Journal of the History of Ideas* **49**, p. 549 (1988).
[79] Cudworth, *True Intellectual System*, I, p. 519.
[80] Turnbull, *Correspondence of Newton*, III, p. 338.
[81] Cited in Manuel, *Newton Historian*, p. 115. Newton makes a similar point in the *Theologiae Gentilis* where he argues that idolatry began in Egypt with 'annual solemnities in honour of their first king & queen Osiris & Isis'. From thence such practices spread to Greece and other countries 'by the colonies of the Egyptians which were very many & the commerce wch ye nations had with one another'. Yehuda MS 41, fol. 10.
[82] Cited in Manuel, *Newton Historian*, p. 91.
[83] Yehuda MS 41, fol. 9v.

84 Cited in Westfall, ' Newton's *Theologiae Gentilis*', p. 26.
85 *Ibid.*
86 Manuel, *Newton Historian*, pp. 39, 93.
87 *Ibid.*, p. 91.
88 *Ibid.*, p. 89 and Wilcox, D. J. *The Measure of Times Past. Pre-Newtonian Chronologies and the Rhetoric of Relative Time*, Chicago (1987), p. 209.
89 *Chronology of the Ancient Kingdoms Amended* in I. Newton, *Opera Omnia* (ed.) S. Horsley, p. 14.
90 Westfall 'Newton's *Theologiae Gentilis*', p. 23 and Manuel, *Newton Historian*, pp. 58—9, 140.
91 Hazard, *The European Mind 1680—1715.*
92 Bernal, M. *Black Athena. The Afroasiatic Roots of Classical Civilization.* Vol. 1 *The Fabrication of Ancient Greece 1785—1985* (London, 1987), p. 191. As Bernal points out, his overall thesis — that the Greeks derived much of their learning from the Egyptians — was a commonplace of seventeenth-century scholarship.
93 Manuel, *Newton Historian*, pp. 98, 119.
94 Yehuda MS 41, fol. 11.
95 Brumfitt, J. H. (ed.), *La Philosophie de L'Histoire*, Vol. 59 *The Complete Works of Voltaire* (ed.) T. Besterman *et al.*, Geneva (1969), pp. 16—7, 32—5.
96 Bossuet, J. *An Universal History from the Creation of the World to the Empire of Charlemange* translated James Elphinson, London (1778), p. 154. Such a providentialist view of history also coloured Bossuet's brief survey of modern history in which he focusses on the history of the Christian Church as 'a perpetual miracle, and a shining testimony of the immutability of the counsels of God' (p. 370).
97 Voltaire, *The Philosophy of History*, London (1822), pp. 155, 69, 147—9. On the Jews' debt to Egypt see also Chapter XXII.
98 Stillingfleet, E. *Origines Sacrae: Or a Rational Account of the Grounds of Natural and Revealed Religion* Seventh edition, Cambridge (1702), Preface. (first edition, 1662).
99 Marsham, J. *Canon Chronicus Aegytiacus, Ebraicus, Graecus* ... , Franeker (1696), [ii].
100 Spencer, *De Legibus*, II, p. 655 and Manuel, *Newton Historian*, p. 101; for references to Marsham in the *Theologiae Gentilis* see Yehuda MS 41, fols. 5, 23. On Marsham and his influence see Rossi, *Dark Abyss of Time*, pp. 125—6.
101 Toland, J. *Letters to Serena* ... , London (1704), pp. 26, 39, 40.
102 Tindal, M. *Christianity as Old as Creation*, London (1730), pp. 90, 282.
103 Spencer, *De Legibus Hebraeorum* ... I, pp. 54—61.
104 Waterland, 'The Wisdom of the Ancients ...', *Works* (ed.) Van Mildert, V, p. 14.
105 Cragg, G. R. *Reason and Authority in the Eighteenth Century*, London (1964), pp. 32—3. On Middleton's religious beliefs see Gascoigne, J. *Cambridge in the Age of the Enlightenment. Science, Religion and Politics from the Restoration to the French Revolution*, Cambridge (1989), pp. 138—41.
106 Middleton, C. *A Letter to Dr. Waterland* In Middleton, C. *The Miscellaneous Works* 4 Vols, London (1752), II, pp. 154, 153.
107 Middleton, C. *A Defence of the Letter to Dr. Waterland* In *ibid*, II, p. 216. Among those whom Middleton thought ought to take more account of such findings was

Newton — thus he took him to task for his contention that the Egyptians 'had not even the use of Letters till about Solomon's Reign' (pp. 231—2).

[108] Rossi, *Dark Abyss of Time*, p. 237. The first part of the *Divine Legation* appeared in 1737 and the second in 1741. The third part was published in fragmentary form in 1788 nine years after Warburton's death.

[109] Iversen, E. *The Myth of Egypt and its Hieroglyphs*, Copenhagen (1961), p. 103.

[110] Warburton, W. *The Divine Legation of Moses Demonstrated* 2 Vols, London (1837), II, pp. 150, 183, 138—9.

[111] *Ibid.*, pp. 119, 91.

[112] Warburton, W. *The Works* ed. Bishop Hurd, 7 Vols, London (1788), V, pp. 294, 288, 290.

[113] *Ibid.*, IV, p. 362,

[114] Warburton, *Divine Legation*, II, p. 91.

[115] Evans, A. W. *Warburton and the Warburtonians*, Oxford (1932), p. 22.

[116] Torrey, N. L. *Voltaire and the English Deists*, New Haven (1930), pp. 9, 128, 170—4 and Brumfitt, *La Philosophie de L'Histoire*, pp. 20, 61, 315—6.

[117] For a review of the literature on this subject see Gascoigne, J. 'From Bentley to the Victorians: The Rise and Fall of British Newtonian Natural Theology', *Science in Context* **2**, 219—56 (1988).

[118] Slotkin, J. S. *Readings in Early Anthropology*, London (1965), p. 229.

[119] Robertson, W. *The Works*, (ed.) D. Stewart, London (1837), pp. lii—iii.

[120] *Ibid.*, p. lii.

[121] *Ibid.*, p. liv.

[122] Frei, H. W. *The Eclipse of Biblical Narrative. A Study in Eighteenth and Nineteenth Century Hermeneutics*, New Haven (1974), p. 152.

GARRY W. TROMPF*

ON NEWTONIAN HISTORY

It is easy to pass off Isaac Newton's preoccupations with chronology and history as *divertissements*. The great man himself sometimes wrote of them disparagingly in this way, and a common impression has stuck ever since that his work on 'natural philosophy' was of far greater importance for him than dating ancient reigns and personages or unravelling the symbology of apocalyptic.[1] The sheer bulk of his writing, however, bespeaks otherwise. From the event-filled year of 1666, when the plague and the Great Fire traumatized London, and when the British fleet turned national disaster into an *Annus Mirabilis* by defeating the Dutch admiral de Ruyter (and forcing the ignominious resignation of Cornelius Tromp), twenty-three year old Newton was already dabbling in divinely appointed dates, while also purchasing prisms and pondering falling apples and 'the force of gravity'.[2]

In troubled times, with the spilt blood of civil war still fresh in collective memory, eschatological speculation abounded, with its attendant mystical numerology and decipherment of apocalyptic images. Millennialist diatribe concerning the four world empires had not died down with the collapse of the Protectorate (in 1659) or with the vociferations of the 'Fifth Monarchy Men' (and their uprisings of 1657 and 1661).[3] There were Anglican divines keeping anti-Catholic interpretations of Daniel and Revelation alive during Restoration years, and nowhere more actively then at Cambridge.[4] During his Cantabrian undergraduate days (1660—4) Newton had procured van Sleidan's *Key to Historie* (or 'Four Monarchies' as he wrote of it),[5] and a Fellow at Trinity could hardly escape the discussions centred on Joseph Mede's extraordinarily influential *Key of the Revelation* (Lat. 1627, Eng. 1643), which had been republished and enlarged in 1664, and which through identifying the Papacy with Anti-Christ in the Apocalypse, re-acquired its earlier political significance when Prince James, brother to the heirless Charles II, publicly disclosed his Catholicism in 1673.[6]

When Newton came to fill the Lucasian Chair of Mathematics in 1669, moreover, he had experienced a good half dozen years of collegiality with the chair's previous incumbent, the colourful and polymathic Isaac Barrow, who had come to mathematics and Euclidean geometry via his study of historical chronology.[7] Those were the years in which the chronographic placement of Biblical and classical events was actually a subject of intense interest and heated debate — even to the laity. The famed (to us somewhat notorious) James Ussher, formerly Archbishop of Armargh, more lately Preacher at Lincoln's Inn and a man not above uttering prophecy about persecuting papists and Rome as the Whore of Babylon, had not long before published his awesome *Annals* (first in Latin, 1650, and then in English, 1658),[8] and entered a field being broached by such doubtly local contenders as Sir John Marsham, Bishop Edward Stillingfleet and Henry Dodwell.[9] In general, mind you, these English chronologists felt in concord as champions of a Protestant prerogative over cosmic time. Despite Ussher's oecumenical-looking prefatorial nod at the labours of the Jesuit Denis Petau,[10] the Protestant and classicist stamp of his work stood as a challenge to mediaevalist and Catholic orientations across the Channel.[11] Newton the young Don breathed the atmosphere of such chronological disputation. From Barrow's study of Ptolemy's *Almagest* and Joseph Scaliger's reassessment of Eusebius' *Chronicon*[12] he would have quickly realized how claims to accuracy depended on astronomical calculations. It was Scaliger's formulation of the 'Julian Period', for instance, which allowed Ussher to pinpoint the creation to that particular Sunday in 4004 BC which 'came nearest the Autumnal Equinox'.[13] And Newton would have sensed how chronography could be as crucial as the interpretation of Biblical apocalyptic for vindicating the truth of his Protestant and Christian inheritance.

Now however difficult it is even for twentieth century scholarship to be exact about when Newton began his first project on matters 'ancient and original', one may fairly affirm that his researches into the Biblical, Pagan and Patristic sources of Antiquity commenced with the opening of his academic career, from thence proceeding *pari passu* with his remarkable investigations into light, calculus, gravity, motion, mass, centrifugal force and the nature of planetary bodies. Concern that he might have to be expelled from Trinity unless he was ordained to the Anglican clergy made him plunge into Patristics in the early seventies. Despite the royal dispensation allowing him to stay (1672), most of his 'silent' decade, — jolted towards its end by the appearance of 'the great comet' sighted over King's college chapel around half-past five in the morning a few days before Christmas 1680! — was spent on Biblical and church history, and sleepless nights at 'Chymical Experiments' to transmute metals.[14] Out of his restless toil with tomes and crucibles there congealed two fundamental and related *principia*, and without grasping either or both one cannot hope to understand Newton's approach to historical and human (as against nature and cosmic) conditions. The one is his covert anti-trinitarian Arianism, the other his equally secretive espousal of the Noachian precepts.

Furtively, at first, in the so-called Common Place Book, he tried to make sense of the debate over the Logos in the theologies of Arius and the homoousian Athanasius (AD 323—36), and critically analyzed some of the proof texts for Trinitarianism in the Bible, including Philippians 2:6—10 and Jerome's Latin corruption of 1 John 5:7—8 (which turned spirit, water and blood into Father, the Word and the Holy Spirit).[15] Cross-checking available new Testament manuscripts, evaluating various Biblical and Patristic authorities on the status of the Son,[16] and then carefully investigating the life and times of Athanasius, Newton eventually arrived by 1675 at his rather original reformulation of Arianism, in which the worship of Christ as God was a basic sin of idolatry and a contamination in Christianity since the fourth century.[17] Just over three hundred years before Timothy Barnes began arguing that Athanasius, a hero of orthodoxy, was some kind of ecclesiastical thug, Newton was tracking down apparent dissembling and inconsistencies in the great Alexandrian's defence against serious criminal-looking accusations at the Councils of Tyre, Sardica and Jerusalem (AD 335, 343—4, 348).[18] Christ, to be certain, was the 'viceroy' of God's Creation, was 'ye word it self ... made flesh', who suffered and mediated for humanity as 'ye son of God', and was exalted to 'honour & dominion after his death', but he was not God.[19] Newton's blood could boil at the apostasy of the Athanasian 'cult of three equal gods'; it was a 'whole fornication' as well as idolatry, and out of the monasticism inspired by the enemy of truth sprang the cult of miracles and 'such other heathen superstitions' as had bedevilled the Church since late Antiquity.[20]

If this were not enough to deprive Newton of his Lucasian chair, after its having been occupied by such a stalwart protagonist for Trinitarianism as Barrow,[21] how curious and theologically dangerous it was that this newborn heir to Socinus, who at no time thought of renouncing his worship of the Christ,[22] should put so many of his eggs into the primitive basket of Noah. The situation is now well known enough, through clear expositions by Richard Westfall and (in this very volume) by John Gascoigne. Around 1672, on Westfall's reckoning, Newton commenced a treatise which he never managed to complete, yet to which he gave a good deal of attention over the following two decades.[23] This was the 'Theologiae Gentilis Orgines Philosophicae' (hereafter 'Origines'), which still sits in its untidy state as part of the Yahuda manuscript collection in Jerusalem.[24] This treatise, while it propelled Newton to immerse himself in the pagan classics over and above his reading of the Fathers,[25] bears a fascinating comparability to his first excursions into church and doctrinal history. The work is about the origins of idolatry (in our post-diluvian world). And its radicalism consists in postulating 'the first religion', that of Noah, as 'the most rational of all others', before 'the nations corrupted it' by worshipping images, and by turning their ancestors into deities as stars, special places, 'hieroglyphical figures', animals, 'and finally into statues and sculptures of all kinds'.[26] Seemingly unique in the seventeenth century, certainly a capstone to his unitarian unorthodoxy, was his conclusion from this that, in essence, the religions of Noah and Christ were the same, and indeed the same as that early apostolic faith which had been corrupted by the evil genius Athanasius.[27]

Why, then, did Newton so universalize the Christian experience of the first four centuries that he denied it any unique role in history? In a century when intellectuals felt more challenged than ever before to explain the world's great, unforeseen ethnographic diversity, with reports coming off the press about the Americas and still more on the inhabitants of Asia,[28] the quest for both what was fundamental in beliefs and practices of the Earth's people and how they look their shape was being taken up in earnest. In Britain this search had been a concern of Raleigh the world historian at the beginning of the century, of Purchas the apologist not long after, and of course deists and Christian Platonists from Lord Herbert to Cudworth,[29] yet few (one thinks here of Hobbes) could view it as a secular enterprise. The Bible did impinge upon it much too greatly. Should Newton be accounted among those so rebellious toward orthodoxy, and did he so thin out the past into a series of oscillations between prophetic enunciations of the one 'most rational religion' and defections from it that he becomes a classic example of developing secular consciousness?[30] For on various leaves found in the Keynes manuscript collection, Cambridge, bits and pieces which may have been intended for the 'Origines', Newton reduces Noah's religion to the two precepts of loving God and one's neighbour, and wrote that 'all the reformations of religion, of Noah, Abraham, Moses, the Jewish prophets and Jesus, are restorations of the oldest religion in the world' — the venerable Noah's.[31]

Why Noah? Of course Noah had not been neglected in the speculative macrohistories and early modern anthropologies of Newton's predecessors. I have already commented elsewhere on that erudite and unorthodox Jesuit, cabbalist, traveller and early modern feminist Guillaume Postel (1510—1581), who like Newton paid such little attention to Adam, and who pictured history running from the Golden Age of Noah to an approaching restoration of that pure pre-Abrahamic Noachian order in the near future.[32] Some of Postel's style of scholarship — especially the idealization of immediate post-Diluvian conditions with a rapid repopulation of the earth, and the revival of Lactantian Euhemerism which supposed that Noah, the 'great mother' his wife, and their children came to be deified as the twelve main divinities of the Graeco- Roman pantheon — seeped into seventeenth-century chronography, comparative mythology and *belles-lettres*.[33] In English writing the Biblical account of Noah was obviously important in confirming the monogenetic origins of all known peoples (Raleigh), and by the end of the century in countering the newly disturbing skepticism which presumed some groups had escaped the flood.[34] Samuel Shuckford (*ca.* 1694—1754) was one of Newton's younger contemporaries who accentuated the slowness of the post-diluvian dispersion along with rampant population growth.[35] Others again pondered the linguistic unity of humanity before Babel, and conjectured in most cases that Hebrew, or in a few opinions Chinese, eluded the confusion and preserved the pristine Noachian tongue.[36] Only a few, Robert Jenkin notable among them, made as much as Newton about Noah's 'true notion of God', which for a time, before its corruption, spread through 'the several parts of the world'.[37]

Yet Newton's focus was rather more specific — on moral precepts — and if the truth be known his key sources were distinctly Judaic. In attempting to construct what the 'Origines' might look like as a complete work, he set down a number of chapter headings, the last of which ran: 'What was the true religion of the Sons of Noah before it began to be corrupted by the worship of false Gods'.[38] Alas, he did not come to the point of writing this chapter, and it takes some detective work to construct what might have gone into it.

First, what were the Noachian precepts, and what of Newton's source(s) for them? In the Tractate Sanhedrin in the Babylonian Talmud, seven of them were listed as commanded to Noah's sons: the institutions of social laws (or councils, or courts of justice), followed by the requirements to refrain from blasphemy, idolatry, adultery, murder (or violence involving bloodshed), robbery, and the eating of flesh cut from a living animal. The Talmudic presumption is that these are the basic regulations to be kept by all Gentile societies, and with prohibitions against sorcery and the making of 'forbidden mixtures' being discussed as possible addenda.[39] The precepts have long been a matter of discussion in Jewry (the first expectations of Gentile proselytes being that they obey them), and they were commented on by the mediaeval Maimonides, a favourite among seventeenth-century European scholars who sought to plumb the depths of Judaism.[40]

Now did Newton ever refer to these precepts in detail? Mysteriously enough, it appears, only once. In a handwritten scholium to his personal copy of the 1717 edition of the *Opticks* (now in the Babson Institute Library, Massachussetts), he formulated a new conclusion to this work by stating that the worship of the sun, the moon and dead heroes among the Greeks and belief in the transmigration of souls, betrayed the original laws of 'our true Author and Benefactor'.

For the seven Precepts of the Noachides were originally the moral Law of all nations; & the first of them was to have but one supreme Lord God & not to alienate his worship, the second was not to profane his name, & the rest were to abstain from blood or homicide & from fornication, (that is from incest, adultery & all unlawfull lusts,) & from theft & all injuries, & to be merciful even to bruit beasts, & to set up magistrates for putting these laws in execution. Whence came the moral philosophy of the ancient Greeks.[41]

Forty-five or more years after his first intentions to write a concluding chapter on these precepts in the 'Origines', then, Newton actually listed them. That is a curious datum to be pondered, considering how personally committed he was to that thread of moral and spiritual truth running from the post-diluvian situation to the early Church. The very unorthodoxy of his position might have been sufficient to explain his reluctance to make a listing, yet however Jewish — even Muslim-looking — his view of the 'essential religion' might have seemed to his contemporaries,[42] all of his most central theological tenets lay hidden from others' scrutiny until his death. With a MS he called 'Irenicum' Newton returned to the old theme of the 'Origines' and the Noachian ethical principles in his declining years (*ca.* 1718). His friend John Conduitt sensed the 'Irenicum' was 'his creed',[43] yet although the projected new ending of *Opticks* written at this time brought him close to showing his hand, he clearly lacked the courage to publish his personal convictions.

Intriguingly, the order of the list of precepts in the *Opticks* scholium does not follow that in the Talmud. It seems like a paraphrase, in fact, of the arrangement in a great Latin work by John Spencer (1630—90), one of the most brilliant Biblical scholars of the age (see previous chapter). In a short *dissertatio* at the end of Spencer's *De legibus Hebraeorum* (1685), a huge work assessing the degree of Egyptian influences on Israelite ritual institutions, Spencer raises the fascinating question as to whether the seven Noachian precepts found in the Talmud (and listed in the order followed by Newton in our previous quotation) were known to the early Christians. The regulatory expectations which the Jewish Christians had of Gentile converts at the Council of Jerusalem (Acts 15:20—1, cf. 21:25) suggested to him that they were.[44]

Newton is not likely to have read Spencer's treatise before 1689,[45] yet it is obvious that its effect was to confirm his earlier instincts, firmly re-stated in the opening paragraph of 'Irenicum', that 'the religion of Noah', which was 'the religion of loving God and our neightbour', was ancient basis for 'the principal part of the religion of Christians'[46] This, however, was only confirmation. The later usage of Spencer's list from the *dissertatio* is only refinement. Newton had apparently acquainted himself with Talmudic materials long before — on preparing the 'Origines'. Curiously the first in a list of ancient sources provided by van Sleidan for budding historians was the Talmud; in Newton's Common Place Book (begun 1660?) we find bishop John Lightfoot's *Horae Hebraicae et Talmudicae* (1658) entered beside a small number of Jewish sources; and we find Newton with a good working knowledge of the 'Talmudists' by the late 1670s.[47]

In a remarkable way, the fact that the last chapter of the 'Origines' was designed to expound Noachide religion, leads us to the wellspring of Newton's whole historiographical and chronologic enterprise. Whether one takes Sanhedrin's Mishnaic presentation of it as the first precept or Spencer's listing of it last, crucial is the requirement of Noah's descendants to appoint councils — *ut Judices crearent qui justa praecepta jam dicta jus administrarent*, as Spencer has it.[48] Why importance can be attached to this dictum is largely because, some time in the late 1680s, while still struggling with the 'Origines', Newton began another work about the beginning of beliefs and institutions, which has come down to us in a well preserved English manuscript at King's College, Cambridge (Keynes MS. 146) under the title 'The Original of Monarchies'.[49] Significantly, this treatise begins with Noah and his sons, and concerns what happened politically once 'ye whole earth was distributed into dependent & coordinate nations tribes & families'.[50] And, despite its title, the real interest of this work, as much for Newton as ourselves, lies in what is said about the instituting of Councils.

Now if the 'Origines' focussed on the beginnings of religious and philosophical notions among pagans, basically to show how idolatory and polytheism emerged as a debasement of Noachian truth, the 'Original' is about the primacy of ancient conciliar government before its undermining by kings. The one work presupposed the other. A short chapter 11 of the 'Origines', part of a short English and thus later inclusion (belonging to the late 1680s) is entitled 'Of the Original of Kingdoms', and stands as the prototype for the 'Original of Monarchies' as we possess it in the Keynes MS;[51] while the latter 'Original', in beginning with Noah, suitably follows what Newton anticipated to be his last chapter of the 'Origines' — on the Noachian precepts. The earlier treatise was on the corruption of these precepts in religion, and the latter on their subversion politically.

Overall, the two works reflect major themes in common, and a wrestling with the same historical problems. The first emphasis in each is that the earliest polities were very scattered and small, and Newton remains obviously reluctant to allow the flourishing of any genuine civilization before the rise of Israel. That was to adopt an age-old stance of Judaeo-Christian apologetics, like Philo Judaeus, Josephus, Justin Martyr, Clement of Alexandria and the rest of them proving that Moses long preceded Lycurgus or Plato.[52] But there were new academic positions to tackle — not the least of which was an increasing conviction among scholars that Egypt possessed a much better claim to be the most ancient and much more of a founding civilization than hitherto imagined — and thus Newton's apologetics for Israel and the ancient mediators of sacred truth took on their own (early) modern flavour. If Josephus had once maintained, for instance, that there was no cohesion in the Egyptians' customs before the arrival of Abram, who delivered to them the science of astronomy[53] (and Newton would have concurred by arguing that all the arts and sciences of civilization sprang from Israel and its satellites), the 'Origines' contains the additional view that Egypt had no political unity before Joseph and the influx of the Shepherds or Hyksos (despite suggestions made by Marsham to the contrary), and the 'Original' only accepts 'the great antiquity of the Kingdom of Egypt' with the severe caveat that the kingdom subjecting the Israelites to bondage was but 'a small part of Egypt', which 'then consisted of several small kingdoms'.[54]

In both these treatises Newton was very much concerned to explain how older beliefs and institutions were diffused to Greece and Rome, and in both much is made of his theses that the Graeco-Roman pantheons derived especially from Egyptian (yet to a lesser extent Mesopotamian) sources. Political and cultural forms in the northern Mediterranean derive from the south and east; it was the Egyptian Cecrops who first brought kingship to Greece, for example, through his rule over Athens; it was a representative of the cultural sphere around Israel — the Phoenician Cadmus of Tyre — who first brought 'the arts' (which include scientific knowledge) to the Greeks.[55] The still more major theme of the works, though, which underlies this account of diffusion, is that of corruption and its spread, Egypt being singled out as the worst multiplier of gods, the chief centre of idolatry and the starting-point of astronomy's counterfeit — astrology.[56] The inceptions of both pagan philosophy and monarchy turn out to be perversions of a pristine, original ordering of life. In contrast — the ancient (Noachian) truths always had a better chance of being preserved when within reach of Israel's light — as with wellsprings of learning among the Ishmaelites (those other sons of Abraham), the Ethiopians, and the Tyrians (whence came Cadmus).[57] Canaan also, Newton would remind us, produced its Melchizedek.[58]

In forging his case Newton makes much use of various lines of chronologic and mythographic lines of thought before him, though impressing his own stamp on folio after folio nonetheless. Throughout the 'Origenes' he was the typical Euhemerist of his time: the pagan gods were really deified humans, and he was only too willing to follow the Lactantian line, restated in Postel, Kircher and Vossius and becoming popular in England at the time,[59] that the twelve primary Graeco-Roman gods were none other than Noah's family and its most impressive descendants, divinization being especially generated along the lineage of Ham, Cush and Nimrod, which resulted in the Assyrian

Empire.⁶⁰ Noah was deified into Saturn in Egypt, and also into Janus after he had fled from his rebellious sons to Italy, while the names Jove and Jupiter, corruptions of the divine tetragrammaton YHWH, came to be applied to people, such as the Assyrian King Belus.⁶¹ Yet Newton was forever wrestling with the tension between the emphases in ancient Patristic chronographers on the temporal priority of Chaldaean civilization and the newer interests in Egypt.

It was the great doyen of Christian chronography, Eusebius of Caesarea (with Jerome and Augustine following him), who placed the first great civilization under Ninus or Nimrod the Assyrian, the founder of Ur (the city of Abram), the son of King Belus, and husband to the fabled Semiramis, who was credited with building Babylon.⁶² The ploy Newton used to alleviate the tension between this older, well-established view and recent claims for Egyptian civilization was first to show how Nimrod's power extended to centres in Egypt and Libya, with beliefs being mediated from Assyria to north Africa via the Arabian Nabataeans (located here far too early!), and then to stress that, in the generations succeeding Noah, there were really only scattered cities and no empires, properly speaking, at all.⁶³

In this fashion, Egypt could be singled out as the earliest source of serious theological confusion. Radically re-adapting Lactantian Age theory, Newton maintained that the passing of four generations after the Flood — those of Noah (the Golden Age), his sons (Silver), Nimrod's time (Bronze) and then the beginning of the Iron Age — saw the basic foundations of ancient polytheism firmly laid in Egypt. In Egypt there occurred a runaway worship of the dead, even of beasts, and Newton could appeal to ancient sources, including Lactantius, that the Egyptians were the first to worship the stars.⁶⁴ The Nile valley, however, was not the sole source of Greek and Roman paganism. Newton was no Kircher, subscribing to an Egypto-centric theory of diffusionism — in advance of Grafton Elliott Smith and John Perry.⁶⁵ There were enough signs that some deities, such as Venus and Aphrodite, could best be explained by looking eastward, to the generation of the Astarte/Anath cult in Assyria out of the unsurpassable beauty of Semiramis.⁶⁶ The figure of Cadmus also stands as the key reminder that not all influences coming to the northern Mediterranean were utterly dark and corrupt.

The general thesis of 'the Original', by comparison, is that the earliest kings were the appointees of councils. The 'first cities' needed captains of their forces against enemies, and chose 'the most honourable & potent amongst their elders' to be King.[67] If this seems like a Polybian gambit, we learn nothing of any other constitutional changes but conciliar governments turning into kingships over a wide geographical area, with the larger monarchy of Assyria eventually deriving from many small city kingdoms, and only emerging at the time of the Hebrew prophets.[68] The earliest Greek cities were run by magistrates until the Egyptian Cecrops founded the Athenian kingdom, and the 'diverse kingdoms in Italy' before Romulus were really ruled by 'ye young men of best note' on behalf of councils.[69] The only serious problems Newton perceived for this whole argument were presented by the exceptional size of Babylon and 'the great antiquity of the Kingdom of Egypt' (see above); but the former remained, after all, a city, and the latter, starting with the nomes that governed its small cities, 'by degrees grew into one Monarchy' since 'the days of Joseph' and 'before the days of Solomon'.[70]

It is only by implication in 'The Original' that the great kingdoms are inferior, or less reflective of the Noachian order, than conciliar governments, although this implication is strong enough when we learn that 'the four Monarchies' (starting with Assyria) rise out of the 'little kingdoms' of 'the first ages'.[71] These four are none other than the tyrannical-looking regimes depicted as beasts in the book of Daniel, rising and falling after each other between the Assyrian empire and the Last Times.[72] All is not decay and darkness in 'the Original', however, because the prytenea, by which Newton means temples for public worship and the maintenance of the sacred fire, continue something of the Noachian order. The earlier, nobler kings were priests as well, acting for councils, and Newton evidently took the preservation of 'ye Vestal fire' to derive from Noah's pleasant-smelling sacrifice after the Flood to avert God's anger towards life on earth forever (Gen 8:20–22).[73]

Newton's methods and historiographical results in these two treatises are very decidedly cast in the mould of seventeenth-century Christian scholarship. His Euhemeristic and etymological procedures will often appear to us quite arbitrary, and they remain difficult to comprehend unless we discern how he was operating in a popular hermeneutical vogue of the time, and also playing off competing applications of Euhemerism found in his Patristic (and thus late antique) sources.[74] Different drafts of the 'Origines' show him toying with a small host of minor theories and thus falling susceptible to self-contradiction.[75] Many of his deductions about the origins of religion and political institutions, moreover, look facile by today's standards. No Champollian had yet arrived to decipher hieroglyphics, and what Newton could make of them mainly served to confirm Egyptian animal worship, with only occasional glimmerings of truth about the universe.[76] Ur of the Chaldees was over two centuries away from the spades of Sir Leonard Woolley, and Sumerian civilization was entirely unknown. These limitations of Newton's time conceded, there was good common sense in stressing the smallness of early socio-political groupings, and his suggestions that Egypt had the most cultural impact on the Greeks are in keeping with new theories about migration and diffusionism around the Mediterranean.[77] His 'anthropology' (or account of humanity's primitive beginnings) was uncluttered with details about a struggle with nature (as with a Diodorus Siculus or Lucretius), let alone about ferocious savagery eventually yielding to reason (as in Polybius and his favourite author Eusebius).[78] It was restrained for concentrating on the post-diluvian situation, in which law and society, arts and skills, could be assumed to have been passed on by the Noachides.

On further consideration, it is highly intriguing and of still greater interest how Newton's pictures of the distant past matched both his ongoing researches into natural philosophy during the last three decades of the seventeenth century and his first serious involvement in English politics. In his work in physics, nothing establishes the linkage better than the one and only glowing emotionally-charged description of a discovery he ever permitted himself, that of the spectrum.[79] As the young Lucasian professor perfecting his lenses for the reflecting telescope, or setting up his *experimentum crucis* to prove that colour was a basic property of sunlight,[80] Newton was uncovering the secrets of the rainbow. The rainbow expressed the Noachian order *par excellence*, it was set in the sky as God's covenental sign for Noah and his descendants to confirm that no similar cosmic catastrophe would ever again befall the Earth (Gen. 9:13—6).[81] The sun, the multi-coloured orb without which no beauty or visual diversity, let alone cosmic bow, could be seen, was also the main source of gravitational pull in our universe, for which Newton had offered a mathematical proof in his *Principia* of 1685.[82] Of all aspects of the physical universe which weighed most for divine providence and most against a universe arising from 'blind chance', it 'was light and ... its refraction', fitting 'the eyes of all creatures after the most curious manner'.[83] When pondering

Newton's interest in the prytenea, moreover, it should hardly escape our attention that fire was the emitter of light, microcosmically related to the fiery orb of the sun itself.[84] Newton, of course, did not anticipate electric bulbs (with yellow light), neon signs or nuclear-powered lighting!

One cannot be certain, but Newton apparently harboured unorthodox cosmological views about the universe passing from one state to another through great macro-periods, with the cosmos being cyclically reduced to the fiery Conflagration referred to in Stoic and apocalyptic literature. Such was the impression he left with John Conduitt during conversations about eschatological matters in Newton's later years (1725),[85] and the views disclosed in this context are not unlike those presented by the first systematic Christian theologian, Origen of Alexandria. In a work significantly entitled *De Principiis*, Origen argued that the coming dissolution of their present world should be understood as the beginning of another (and preferable) cosmic order, which would in turn be succeeded by another.[86] It is just possible that Newton believed the present shape of the Earth was due to a shift of cosmic principles at the time of the Flood, which was a catastrophe paralleled to eschatological events in the Bible (e.g., II Pet. 2:4—9, cf. Matt. 24:37, Lk. 17:26). During the 1680s he was maintaining that the pendulum would be found to swing more slowly nearer the equator, a calculation confirmed empirically by continental investigators in the Tropics. If the rotating Earth were in the simpler shape of a ball, he summised, without 'flattened' poles and 'swollen' equatorial regions, its vast watery surfaces would be out of control.[87] Counting against this, perhaps, is the deduction by optical experiments that the sun itself was not a 'perfect sphere',[88] and in any case we lack Newton's comments on the physics of the Deluge. We do not know if he would have concurred with his friend Thomas Burnet that the Deluge followed the condensation of vapours released from the Earth's centre by a gigantic earthquake, or with his would-be protégé William Whiston, that the passing of Halley's comet drew up subterranean waters which were normally pulled closer to the Earth's centre by gravity.[89] That he saw his work in natural science very much concomitant with his view of Noah as the custodian of essential religion and wisdom for present humanity, however, is a fair deduction. Why, even his youthful interest in a universal language bespeaks a concern for the loss of humanity's unity after the Flood (Gen 11:1—9), while his fascinations with the ancient lineal measurement of the cubit, which resulted in a (posthumously published) treatise during the 1680s, have their background in the proportions of Noah's Ark as recounted in Gen 6:15—6.[90]

A comparable interest lies in the relationship between Newton's historical research and political positioning. In 1687—9 he sided with exclusionists in disallowing Catholics, and in particular the Benedictine monk Alban Francis, from receiving his Masters degree from Cambridge. Quite apart from his covert Arianism, Newton naturally sided with a common anti-Catholic cause to block the infiltration of Papist idolaters (who were also members of the tradition most critical of his scientific work).[91] Abetting adverse Protestant reaction to James II's unsanctioned Catholic policies, and for a short time fully involved in the politics of the Glorious Revolution (as a Member of the Convention Parliament 1689—90),[92] Newton was without question a 'Parliamentarian' in sympathy. 'The Original of Monarchies', by implication, legitimates kingship only insofar as it is responsible to conciliar — we may translate 'parliamentarian' — government. Forged as it was in years of turbulence and transition (*ca.* 1687—93), this treatise also seems to gather up strands of Presbyterianism which go back to the Puritan John Angell and other boyhood teachers at Grantham School, Lincolnshire, during the Civil War.[93] Although, as we shall see, Newton was to make much of the great kingdom of Solomon in ancient Israel, nothing clinched the mandate for parliamentary limitation of monarchial power more than the primordial injunction among the Noachian precepts to establish Councils, as well as its upholding through Antiquity in the Israelite judicial and Synagogal system.[94] Like James Harrington and not a few Puritan theorists from the mid-century, he treated Israel's Biblical institutions as models for English constitutional rectitude.[95] Even Oliver Cromwell's most prominent political icon, it should be recalled, was the ark at rest under the directed and demarcated rays of the sun,[96] while Newton had nothing from his researches into the ancient past to justify the royal benevolence of a Sun King or the monarchical authoritarianism of a Hobbes.[97] Excepting, that is, the reign of Solomon.

Solomon was a crucial linchpin in Newtonian historiography. To some extent this was so by an inheritance from contemporary chronography, for had not the estimable Ussher fixed the opening of the great Solomonic temple at 1004 BC, exactly three thousand years after the Creation?[98] Newton, however, was not interested in chronological symmetry so much as accuracy; timing the Temple's inauguration, which he placed in the apparently insignificant year of 1015 BC,[99] was of far less importance than establishing inviolable principles for a synchronous history of the ancient nations. Significantly, he did not consider such synchronicity possible for events occurring more than a century before the reigns of David and Solomon.[100] Already in the 'Origines' there is a reluctance to put dates on reigns in Chaldaea or Egypt even after the fashion of Eusebius' (and Jerome's) *Chronica*; and his only attempt at chronographic precision in that treatise was to deduce the Creation to be 2400 years later than the Chaldaean Berossus had claimed in Antiquity.[101] Already in 'The Original', moreover, Newton strongly suspects that pagan historians 'raise their antiquities higher than truth by many thousands of years'.[102] This is precisely the pretext for writing his intriguing *Chronology of ancient Kingdoms amended*, both in its abridged and longer versions.[103] When we read at the conclusion of 'The Original' how he has corrected impressions about the limited extent of Egyptian growth in earlier times, and how the genuine unification of Egypt 'remain's now to be explained',[104] we begin to perceive that Newton was engaged in a great and continuous work of historiography. For it is a fundamental argument of the *Chronology* that all the great monarchies of the ancient world only achieve civilization and greater unity in the shadow of Solomonic civilization (and the proximate kingdom of Phoenicia or Tyre),[105] and Newton's notes show him preparing for this case in earnest by 1699.[106]

The real trouble facing apologists for Biblical priority in Newton's time was the Hellenistic Egyptian Manetho, a priest who rendered a complete listing of the pharoahs, along with (generally antipathetic) asides on early Israelite affairs (which were linked by him to the Hyksos). Although used with confidence by patristic writers, and although being of little interest to Renaissance humanists, Manetho came to provide seventeenth century scholars with fresh reasons for revising Egyptian chronology.[107] If esotericists' claims about the great antiquity of the Egyptian Hermetic literature had proved faulty (by 1614),[108] the threat of Manetho's list remained, with its suggestion of more than ten millennia of greatness before Israel. Newton joined others in arguing that many of the reigns should be taken as synchronous rather than successive, and that Manetho did little else but exaggerate when setting them in elongated dynastic lineages.[109] And

Manetho had not been alone; unlike the trustworthy Biblical record, Greek chronologies, for one other glaring case, had unduly lengthened human generations and the age of Spartan kings.[110]

Now despite all the curiosities and varied particulars of the Newtonian Chronicles (covering the years 1125—331 BC), major apologetic arguments clearly reveal themselves. While there were continuing exercises in speculative Euhemerism — Ceres as a Sicilian teaching Greeks how 'to sow corn' (in 1030), Prometheus freed as a captive in the Caucasus (in 937), Theseus killing the Minotaur and Dido founding Carthage in the precisely denoted years of 968 and 883 respectively, and so forth[111] — most important is the message that Egypt had no great empire until Sesac (= the Biblical Shishak and the classical Sesostris), beginning from the day's of Solomon's successor Rehoboam (in 974), and that arts were brought into Europe principally by the Phoenicians and Curetes, in the time of Cadmus and David, about 1041 years before Christ, even though such arts were known in Israel's sphere — in Phoenicia, Egypt, and Idumea — before that.[112] By thoroughly historicizing Sesostris' fabulous (clearly mythical) sway over India and most of Asia, and over Thrace and north Africa (to the years 974—56), Newton made up for his relative denigration of earlier Egyptian achievements.[113] If anything convinced him of the value of his own Euhemeristic approach more, it was the case of Sesostris, who was well known in Hellenistic sources as a person who once lived and became divinized. In his lifetime, Newton argues, the worship of the Pharoah as God begins. Sesostris expected to be treated as Osiris, and his queen as Isis, and at death he became Bacchus for the Arabs, Mars for the Phrygians and Hercules for the Greeks. In all this the very opposite to Solomon's righteous rule is made manifest, and besides, Sesostris was a despoliater of Jerusalem's Temple.[114]

As for Cadmus, he takes on a twofold cultural significance. First, it was only his coming from Tyre to Hellas (in 1045 BC) which makes possible the writing of any reliable history among the Greeks (and Latins).[115] Newton satisfied himself that the Bible and the satellite historical tradition of Phoenicia (known especially through Josephus and Christian Fathers Tatian and Theophilus of Antioch)[116] provided the only chronologically accurate or unexaggerated fashion of recording past events before members of other cultures (such as Herodotus) tried their hand at historiography. Second, since Cadmus bore Phoenician 'sciences', not only writing and 'other arts ... and customs' to

Greece,[117] he paved the way for the earliest (known) accurate observations of the heavens. The fact that the general ('Aramaean') cultural milieu had its impact in the civilizing of Egypt was significant enough, and for anyone who has ever wondered why Newton began his Chronicle with the departure of the Hyksos from the Egyptian arena,[118] it was precisely to intimate this point. But Cadmus, whose movements were conjectured to have some connection with this departure and with population shifts in the Phoenica-Palestinian area,[119] bore the knowledge that could produce an astronomer's sphere. As Frank Manual has expounded in marvellous detail, the earliest date in ancient Greek affairs Newton was most confident in fixing was the Voyage of the Argonauts. Far from being a mythical journey, this voyage was very much facilitated by what Newton believed to be a 'primitive sphere' in the hands of Chiron the Centaur, who (according to Newton's curious rendering of a Greek phrase in Clement of Alexandria) was a teacher of 'asterisms' and would have had to possess such a sphere.[120] This Chiron, Newton hypothesized, would have placed the colures (the projections of two octants of the celestial sphere on to the sphere of the fixed stars in the shape of two adjacent spherical triangles) in a balanced position (or at the 15th degree), with the line separating these two halves of the Zodiac passing through the signs of Aries, Cancer, Chelae and Capricorn after the 'oldest astronomers'.[121] Since there were precessions of the equinozes at 1° every 72 years, it could be deduced from later Hellenistic measurements of such shifts that a chronological placement of the ancient voyage could be made with unprecedented scientific accuracy — to 939 BC.[122]

This was a date a little over 150 years less ancient than a common reckoning,[123] and allowed Newton to put the events of the Trojan War after Solomon (Troy falling in 904 BC).[124] The real pretext of the Argonautic expedition was the further purveying of the news, first brought by 'ye Merchants of Phenicia [sic]', that the great Egyptian empire created by Sesostris had collapsed.[125] The struggle between the forces of the Greek Agamemnon and Priam of Troy came later. Newton, along with his contemporaries, had no reason to be as skeptical about the historicity of such characters as many nineteenth century liberals to come, but he lacked Schliemann and his successors to provide a more correct (and earlier) chronological placement of Mycenaen times.[126] That may be being unnecessarily charitable. After all, he was only too eager to give the kingdom of Israel pride of place in

Antiquity, and Newtonian history is all about 'finding the evidence' accordingly.[127]

There are various reasons why the Davidic, and especially Solominic, monarchy was the centre-piece in Newton's picture of ancient history. *Inter alia* this kingship was in possession of what were 'by far the oldest records new extant',[128] and (following some Hobbist-looking Biblical criticism) all the narrative and legal records preserved prior to this monarchy's emergence were put into order by Samuel (who was of course the very prophet to anoint David as king).[129] It cannot be stressed too highly that Newton believed the Bible (in its uncorrupted texts) to be divinely authorized, and Deist though some would make him, did hold it to be an account of God's active dealings with humanity.[130] Another datum rendering Israel's kingship immensely significant was the Solomonic Temple. Not only was it a monument combining great civilization with true religion, but its proportions — in sacred cubits again — were invested with mystical significance. Its mathematical secrets betokened both the foundation wisdom of the Noachian order in the past (which was mediated via the Mosaic tabernacle), as well as the descent of the heavenly temple in the eschatological future (cf. Rev 21:2).[131] Its building under Solomon was also connected with his friend King Hiram of Tyre, the Phoenician monarch who supplied materials and skills (1 Kgs. 5:1—12, 18), and whose royal archive both complemented Biblical materials and evoked the 'golden fleece' of esoteric wisdom.[132] Even the great temple pyramid of pharoah Menes is built after it (and thus by implication inspired by it), 'the priests of Egypt' subsequently making its erection 'above a thousand years older' than it was.[133]

The history of astronomy and natural philosophy are not left disengaged from this orientation. Without Cadmus and influences from across the Mediterranean — as Josephus had already implied — 'the consideration of things celestial ..., things which are supposed to be the oldest of all among the Greeks' would have arrived much later.[134] Pythagoras, whom Newton singles out, not only as the founder of Greek science but the best preserver of the true religion, had to be placed much later than Solomonic times in any case (in the sixth century BC with Thales).[135] But Pythagoras was the inheritor of the most ancient (and Noachian) truths — a *prisca sapientia* as McGuire and Rattansi have neatly put it.[136] The coupling of profound cosmological insight with that knowledge of the supreme God which Pythagoras shares with Plato and other Greek philosophers was derived, in Newton's estimates, from Egypt (rather than Chaldaea). In Egypt, he surmised in the most Euhemeristic fashion, there had already been legitimate astronomers such as Thoth, Neptune and Pluto.[137] In the new scholastic (substantially Cartesian) ethos, interestingly, Newton held it to be 'a very ancient opinion' that 'all matter consists of atoms'. Relieving the atomist tradition of its old *odium erroris*, he noted how Pythagoras shared the opinion, which he significantly agreed to have derived 'from Egypt and Phoenicia' via 'Moschus the Phoenician'.[138] And Pythagoras, through grasping the 'music of the spheres', had also recognized God as Harmony and thus the cause of gravity.[139] If these basic ideas had more ancient foundations, however, with Pythagoras himself placed four hundred years after the era of Solomon, Hiram and Moschus, we should now ask ourselves what Newton wrote of developments through the rest of ancient history.

The leading macrohistorical structure in Newton's interpretation of post-Solomonic affairs is that of the four world Monarchies. The first of these is not Assyria, as in certain Patristic frameworks, for Assyria and Egypt stand in 'The Original' as preliminary kingdoms which emerged from scattered cities (see above). The first is the Babylonian empire (together with its contemporary Media), the second Persia, the third, Greece, and the last Rome.[140] The famous Biblical cue for this patterning was provided by the Book of Daniel, with some of its Patristic and latter-day commentators. Hardly any homage was paid to the Augustinian paradigm of the two great Babylons, with the first being founded as Babel and rebuilt by Semiramus yet envisaged to collapse at the very time the second Babylon (= Rome) began its course.[141] Newton inclined to a more undeviating biblicism and he was influenced enough by the methods of Joseph Mede to be persuaded that 'the prophetic-apocalyptic' works of both Daniel and its New Testament counterpart Revelation were historical texts veiled in symbol.[142]

Now the classical historiographical works of the Old Testament offer no account of events beyond the middle of the Persian period (on Newton's reckoning 412 BC),[143] yet the narrative framework of Daniel looks to all intents and purposes historical, and the context of its hero's life is the transition from the Babylonian to the Persian regimes (1:1— 9:1).[144] The Danielic visions first address this great shift from one great empire to another, but project the coming of two more (2:31—45; 7:1— 8:26; 11:1—45). It is in the unravelling of actual historical events indicated by the divinely inspired prophecies of Daniel and the Christian Apocalypse that Newton penned his (posthumously published) *Observations upon the Prophecies*, and thus the last segments of his intended *magnum opus* on Antiquity. That his concerns with apocalyptic go back to the 1660s[145] only goes to confirm the integrated nature of his historico-chronological researches all along.

That Newton intended his *Observations* as a sequel to his *Chronicle* seemed obvious to his friend John Conduitt (*ca.* 1725).[146] Since the interpretations in the *Prophecies* only take the reader to the rise of Islam,[147] moreover, it is palpable that Newton's historical enterprises were consciously limited to Antiquity, and were especially governed by the scope of the sacred Biblical record. Not that this record was without relevance for the future of his own generations, for he was probably tempted to finish the *Prophecies* with comments on the last chapters of Revelation.[148] This would have been to provide but an addendum, however, and with an avoidance of speculation about the timing of the Last Days.[149] His main purpose was to reveal the extraordinary prophetic accuracy of Daniel and Revelation by sticking to precise 'Rules for methodising' the significations of apocalyptic figures and beasts along Mede's lines. '... Historian's divide their histories in Sections, Chapters and Books', Newton contended,

at such periods of time where the less, greater and greatest revolutions begin or end; and to do otherwise would be improper: much more ought we to suppose that the holy Ghost observes this rule accurately in his prophetic dictates.[150]

This may appear to be an unlikely *argumentum ad hominem*, but Newton had recourse to other ancient accounts of events only insofar as they were reflected in scriptural prophecies.

While Newton's great interest in Revelation belongs very much to the temper of the times,[151] he was Patristic scholar enough to know that it was the least acceptable of the New Testament works to the upholders of Nicene orthodoxy. If the doyen of chronography among the Fathers, Eusebius of Caesarea, had questioned its inclusion in the New Testament canon, at least the great Origen, among other pre-Nicene Christian authorities, had made extensive use of it.[152] Revelation, Newton's exegesis apparently has it, refers to Athanasius as the false prophet, in the context of Constantine's rule and the splitting of the two Romes.[153] Athanasius, we remember, is the theological arch-villain by which popish idolatery and 'monkery' are unleashed on Christendom, and it is a false idolatrous Trinitarianism which also produces corruption of other New Testament texts to suit Nicene doctrines.[154]

In Newton's eyes, then, it is precisely the last identifiable historical references in Revelation which prophesy the abandonment of sacred truth. The divine injunction to Noah to form Councils (of Elders or *presbyteroi*) was abandoned for papal monarchy; and the two primarily religious prescriptions among the Noachian precepts — not to alienate his worship and profane his name — were the very commandments betrayed by Romish iconic practice and Trinitarianism. There is little to add here about the other four precepts — to abstain from eating blood, fornication, theft and all injuries (even to beasts) — except to surmise that Newton's refusal to eat black pudding, hatred of cruelty to animals, and his remarkable personal chastity, reflect personal commitments apparently having something to do with them.[155] It is conciliarism and anti-idolatry which are the master themes of Newton's theology and ethics, or the very hub and axel around which his historical excurses revolve.

In what appears to be a motif of historical recurrence, the precepts of Noah were re-enunciated through sacred history — by Abraham, Moses and the Prophets, as well as by Christ and the early Church. The reclamations of 'the true religion' were necessary precisely because there were recurrent defections.[156] How Newton conceptualized each recovery of truth is naturally a matter of some curiosity. After all, does not Moses sit ill in this chain of the greats, for one, since he is associated with an elaboration of law in the Hebrew Bible and with legalism in the New Testament? And how Newton pictured the revolts from truth is of comparable interest, if only because of the suggestion of some long-term connection between ancient and modern Romish idolatries.

Since Newton acknowledged that Jesus' two commands to love God and one's fellow humans were summations of Noah's regulatons,[157] it is possible the thoughtful Englishman held the highly advanced (but in his time unorthodox) view that Moses proclaimed the Ten Commandments only, or else that he meant to distil them in the separated rulings to love God and one's neighbour (Deut. 6:5, Lev. 19:18). On this reading of his views, most of the serious spiritual defections from truth can then be linked to Egypt. According to the 'Origines' and related jottings, it was the Egyptians who corrupted Noah's religion by 'the worship of their dead Kings & Heroes', and although in a sense truth lasted in an attenuated form in Egypt, it only properly 'continued in Abraham & his posterity who revolted not'.[158] Abraham and his posterity, we can infer, are directly linked to the impact on Egypt of the Shepherds (or Hyksos), who 'were of another religion' than the 'preceding kingdom of the lower Egypt'.[159] Only after these Shepherds are 'expelled' do the Egyptians 'set up the worship of their own kings and princes', and thus revolt against truth again.[160] Moses may have 'reclaimed' his truth, but one detects the Spencerian implication in his approach that early Israel accepted enough of Egyptian legal and institutional principles to reproduce subsequent idolatries. These were those challenged by the prophets and eventually Jesus (whose chronological placement, miracles and message were not left without some special attention in Newton's writings).[161] The Israelites also rejected the Messiah, and if this revolt was not in Egypt, it was the Egyptian Athanasius who polluted both the message of Jesus and the (partial) re-statement of the Noachian precepts at the first Council of Jerusalem (AD 49)[162] with his party's Nicene formularies. Far from the notion of the Trinity originating with Moses, moreover, and marvellously reflected in Greek philosophy, as John Wallis, Theophilus Gale and Ralph Cudworth had argued,[163] this doctrine was symptomatic of theological disfigurement.[164] The new Roman idolatry consequent upon the Athanasian error had lasted to Newton's own time, and whether or not he concurred with Joseph Mede that there was a steady retrieval of truth from the Albigensian reaction to Catholicism up until his own time,[165] he secretly assigned a special role for himself as the re-discloser of the same old true religion on the threshold of a new temporal dispensation.[166] Virtually 'venerated as a divinity' in his own time, it was perhaps inevitable that he himself should find a place in his own *Heilsgeschichte*.[167]

The Noachides' age, as history's fountainhead, also bequeathed to humanity occult wisdom. This *philosophia* combined ethical with profound cosmological insights. In certain editions of the *Opticks*, we recall, Newton derived Greek philosophy from Noah's precepts, and all editions, in fact, end with an allusion to pagan Moral Philosophy. Over and above Pythagoras, the 'ethics, or good manners' in the teachings of Socrates and Cicero are reflections of the precepts, as also the moral principles of the Chinese Confucius.[168] But the deepest scientific truths of Antiquity also spring from this connection. The more one ponders Newton's axial principles, in fact, which he divulged in their sevenfold form only once, and then one relates this covert, Talmudically-inspired unorthodoxy to his fascination for the mysterious proportions of the Solomonic temple, the more one can sense the milieu of early Freemasonry. Between its foundations in England (at the hands of such members of the Royal Society as Elias Ashmole and Robert Moray)[169] and its systematization under the Talmudist James Anderson in 1723 (who added more obviously Christian and Rosicrucian principles to its platform),[170] the precise character of Freemasonry remains elusive. That the [Masonic] seven virtues substitute for seven earlier ones is highly likely; that both a concentration on Solomon's temple and a connecting of its secrets to Noah and the 'Hiram legends' go back to a pre-Andersonian set of Masonic interests is a fair deduction. Yet whether Newton was party to these Masonic interests or a critic of them cannot be easily decided.[171] Signs of Newton's reserve may have much more to do with his unorthodoxy and a dread of criticism than dabbling with the occult.[172] On the other hand, they also probably have a good deal to do with his belief that true knowledge and wisdom were occult, or hidden secrets of the divine, to which those of right spirit might be privileged to disclose experimentally.[173]

One obvious conclusion remains, that Newton's natural philosophy and his theologically-oriented approach to history are two closely related strands in the one quest and human *opus*. It may be correct to argue that he tended to keep 'religion' and his scientific endeavour separate, simply because religion, when applied personally and not to the beliefs of the ancients, still connoted 'worship' and 'piety' (as in most seventeenth century usage) rather than any objectified social forms.[174] As for theology, ethics and the interpretation of history both sacred and profane, they were integral and crucial to his pursuit of truth, and thus not to be examined apart from his formulation of physical laws and mathematical axioms. For to him, let us reiterate, Pythagoras was both bearer of ancient mathematical wisdom and the true religion to Europe (which had again been corrupted by polytheistic idolators). In Justin Martyr a quotation from Pythagoras couples the great Samian's belief in one God, with affirmations about him as the One creating all, 'the Light of Heaven, the Father, ... [and] the Mover of all spheres'. Newton conceived himself cast in a similar role.[175]

Who could blame one for the summary assessment, however, that Newtonian history does not conform to the inductive and experimental methods he set himself in the study of nature?[176] For all its fascination, his great historiographical enterprise was agonized out of age-old theological premises about scriptural truth, untestable hypotheses about ancient events and apocalyptic symbology, Euhemeristic manipulation and chronological fudging, and a tendency towards a forced patterning of events that is endemic to most macrohistorians as 'arbiters of meaning.'[177] Paradoxically, his sense of historical time and geographical space left something to be desired; even Stukeley, his somewhat unreliable early biographer, sensed how Sir Isaac 'shortend the years of the world ... a little too much' (though Stukeley had possibly heard of great Nicolas Fréret's strictures), and the newly discovered Americas barely figured in Newton's visualization of the habitable earth.[178] These features only go to confirm that his was no secular mind, and on his reading the past is redolent with a divine ordering. The characters of his history, it has been rightly adjudged, may lack all personal qualities — seeming to move like billiard balls yet on a social stage! — and with the nations as 'neutral as astronomical bodies'.[179] Yet rather than being a reflection of the new 'mechanical philosophy' (which in Newton's estimation lent itself to atheism),[180] his visualization of the past has a gothic organicism about it, with history taking its complex courses under the direct governance of the divine. 'All the truths of God's creation were once revealed' in a primordial situation, what is more, 'as an interconnected whole which comprised natural, moral and divine knowledge'.[181] Beside a Machiavelli or a Bayle, his work bespeaks as much, if not more, of traditionalisms in the past than of liberalisms in the future.[182]

ADDENDUM

At the time of writing, Penelope Gouk's article 'The Harmonic Roots of Newtonian Science', in *Let Newton Be!* (eds. F. Fauvel *et al.*), Oxford, 1988, ch. 5 was unavailable to me. The connections she finds in Newton's optical work between the seven colours of the spectrum and the seven notes of the musical scale confirm his profound concern to rediscover truths known to Pythagoras, as also do Piyo Rattansi's findings in an article from the same collection. Rattansi's biographical study of Newton had not yet been published at the time I tackled this piece.

NOTES

* Thanks are due to my colleague Dr. John Gascoigne (History Department, University

of New South Wales, and currently an Associate of the Department of Religious Studies, University of Sydney) for access to manuscriptal materials and his expert bibliographic advice; to Mrs Ruth Lewin-Broit, my Research Assistant, who was especially helpful with Hebrew sources, to Messrs James Rigney and Stuart Simpson, Rare Books Library, Fisher Library, University of Sydney; Mr Hans Arns of St. Patrick's College Library, Manly, Sydney; and to the custodians of the Bodleian Library and the Museum of the History of Science, University of Oxford, and of the rare books (and microform) collections in the State Libraries of New South Wales and Victoria, and in Moore College, Sydney.

[1] See F. E. Manuel, *Isaac Newton Historian* (hereafter *INH*), Cambridge, Mass. (1963), chs. 1—2.

[2] On dating these early interests, R. S. Westfall, *Never at Rest; a biography of Isaac Newton*, Cambridge (1980), pp. 156—60; Manuel, *A Portrait of Issac Newton*, (hereafter *INP*, Cambridge, Mass. (1968), pp. 78—83; cf. H. Pemberton, *A View of Sir Issac Newton's Philosophy*, London (1728), pref. (whence quotation); D. E. Smith, 'Two Unpublished Documents of Sir Isaac Newton,' in W. J. Greenstreet (ed.), *Isaac Newton, 1642—1727*, London (1927), pp. 16—34. On the plague and the international politics of the day, see e.g., G. M. Trevelyan, *England under the Stuarts*, London (1904) (repr. 1966), ch. 11; J. R. Jones, *Britain and Europe in the Seventeenth Century*, London (1966), chs. 4—5, cf., John Dryden's *Annus Mirabilis* (using *Poetical Works*, (ed.) G. Gillfallan, Edinburgh (1855), Vol. 1, pp. 40ff.).

[3] For background, e.g. J. R. Tanner, *English Constitutional Conflicts of the Seventeenth Century 1603—1689*, Cambridge (1952), chs. 11—13 (more generally), L. F. Brown, *The Political Activities of the Baptists and Fifth Monarchy Men in England during the Interregnum*, Washington (1912) (specific). On the four world empires in early modern European historical reflection, G. W. Trompf, *The Idea of Historical Recurrence in Western Thought*, Berkeley (1979), Vol. 1, esp. p. 344, and here I foreshadow Vol. 2, ch. 1. Cf. also ns. 5, 47 *infra* on Sleidanus.

[4] E. L. Tuveson, *Millennium and Utopia*, New York (1964) edn., pp. 76—87; M. C. Jacob and W. A. Lockwood, 'Political Millenarianism and Burnet's *Sacred Theory*', in *Science Studies* **2**, 226—70, cf. esp. 267 (1972), n. 4 and the contemporary literature cited there. Cf. also R. H. Popkin (ed.), *Millenarianism and Messianism in English Literature 1650—1800 (Clark Library Lectures 1981—1982)*, Leiden (1988).

[5] In Newton's undergraduate reading notes; thus Westfall, *op. cit.*, p. 82, n. 46, cf. J. van Sleiden(us), *Key to Historie* (trans from the Latin edn. of 1559, cf. n. 47 *infra*), London (1631).

[6] Mede, *Clavis Apocalyptica*, London (1627), Eng. as *Key, etc.*, (trans. R. More from the 2nd Latin edn. of 1632), London (1643). On the Catholicism of James II, e.g., J. F. H. New, *Anglican and Puritan*, London (1964), pp. 88—9.

[7] Thus A. Hill, 'Some Account of the Life of Dr. Isaac Barrow', in V. de Sola Pinto (ed.), *English Biography in the Seventeenth Century*, London (1951), p. 152, cf. also J. Gascoigne, 'Barrow's academic *milieu*: Interregnum and Restoration Cambridge', in M. Feingold (ed.), *Before Newton: the life and times of Isaac Barrow*, Cambridge (1990) (forthcoming).

[8] *Annales Veteris et Novi Testamenti, etc.*, London (1650); Eng. as *The Annals of the World. etc.*, London (1658), cf. his *Wonderful Prophecies of these Times* (1689), reprod in *The Whole Prophecies of Scotland, England, Ireland, France and Denmark*, Edinburgh, n.d. (nineteenth century) (prophecies); *An Answer to a Challenge Made by a*

Jesuit in Ireland, Cambridge (1835) edn. used, esp. pp. 3, 430—9 (against papist idolatry).
[9] Marsham, *Canon Chronicus Aegypticus, Ebraicus, Graecus, et disquisitiones*, London (1672); Stillingfleet, *Origines Sacrae*, London (1662); Dodwell, *Dissertationes Cyprianicae*, Oxford (1684) (with important appendices).
[10] *Annals*, p. [ii].
[11] For background J. W. Johnson, 'Chronological Writing: its concepts and development', in *History and Theory*, 2/2, 137 (1962).
[12] Hill, *op. cit.*, p. 152 (on Barrow); A. T. Grafton, 'Joseph Scaliger and Historical Chronology: the rise and fall of a discipline', in *History and Theory*, 14, 156ff (Scaliger), (1975).
[13] Ussher, *op. cit.*, pp. [iv—v]. Ussher also knew Scaliger's computation of the 'Julian Year' from Petau's *De doctrina temporum*, cf. the 1703 Antwerp edn. (ed. J. Harduini), Vol. 1, esp. pp. 87, 417—9. For background to this computation, D. Wilcox, *The Measure of Times Past*, Chicago (1987), pp. 188, 198f., and for a fairly early reference to Scaliger by Newton, see Yahuda MS 16.1 [or var. 1, Newton Papers 16], folio 28^r (1675?) (all Yahuda MSS being from the University of Jerusalem).
[14] For most of the above, Westfall, *op. cit.*, pp. 310, 335—410, cf. p. 393 (first quotation); Humphrey Newton, Keynes MS 135 (all Keynes MSS, at King's College Library, Cambridge), as in *ibid.*, p. 361. See also D. Brewster, *Memoirs of the Life, Writings, and Discoveries of Sir Isaac Newton*, Edinburgh (1855), Vol. 2, pp. 98—101 on the royal dispensation.
[15] As in H. McLachlan (ed.), *Sir Isaac Newton's Theological Manuscripts*, Liverpool (1950), pp. 129—33.
[16] *Ibid.*, p. 133; early 'Notebooks' (Keynes MS 2), esp. 'De Trinitate', pp. 33ff., 'Observations upon Athanasius' Works', pp. 13ff.
[17] See Westfall *op. cit.*, pp. 314—5 and the MS materials cited there.
[18] Yahuda MS 14, f. 83^v, cf. 'Paradoxical Questions, etc.', in McLachlan, *op. cit.*, pp. 68—82; 'The History of the Council of Nice, etc.' (Keynes MS 4), and note T. Barnes, 'The Career of Athanasius', paper delivered to the Tenth International Conference on Patristic Studies, University of Oxford, 26 Aug. 1987 (for *Studia Patristica*, and foreshadowing his monograph on *Athanasius and Constantius*).
[19] Yahuda MS 15.7, f. 154^r (first quotation, cf. also 15.5, ff. 96^r—97^r); Keynes MS 2, p. 15 (other quotations), cf. Yahuda MS 14, fs. 25 (esp. pts. 5—11), 84^v.
[20] Yahuda MS 11, 8.7; 2.2, f. 19; Clark Library MS (Los Angeles) as in Westfall, *op. cit.*, p. 345 (three quotations respectively).
[21] See Barrow's, *A Defence of the Blessed Trinity*, London (1697), and *A Brief Exposition on the Creed*, London (1697), (cf. also *Works*, et. J. Hamilton, London (1844—5), Vol. 2, pt. 1).
[22] E.g., 'Common Place Book', McLachlan, *op. cit.*, pp. 130—1. Cf. G. H. Williams, *The Radical Reformation*, Philadelphia (1962), pp. 749—63; J. H. Colligan, *The Arian Movement in England*, Manchester (1913), on Sociniarism.
[23] Westfall, 'Isaac Newton's *Theologiae Gentilis Origines Philosophicae*', in W. W. Wager (ed.), *The Secular Mind; transformations of faith in modern Europe*, New York (1982), pp. 16—7.

[24] Esp. in Yahuda MS 16, folios 1ʳ—38ʳ to be called 16.1) with redactions apparently being made on the rear of an original MS as from 10ʳ, = on my rendering folios 1ᵛ—79ᵛ (to be called 16.2).
[25] Westfall, *Never at Rest*, esp. p. 353, cf. p. 312; and the manuscriptal evidence cited there.
[26] Yahuda MS 41, f.7 (first two quotations); 17.3, f.8 (last).
[27] Esp. Keynes MS 3, p. 35; Yahuda 16.2 f.45ᵛ, MS 7.2p (surviving scrap of paper), cf. Westfall, '*Theologiae*, etc.', *loc. cit.*, p. 28 (on true religion), and see n. 18 *supra*, cf. Yahuda MS 15.7, f.154ʳ where Arius, along with the chief villain Athanasius, received censure, as Manuel, *The Religion of Isaac Newton* (hereafter *INR*), Oxford (1974), p. 58 has rightly noted.
[28] Cf. G. Atkinson, *Les nouveaux Horizons de la Renaissance Française*, Paris (1935), pp. 10—12, etc.
[29] For surveys of relevant literature, e.g., M. T. Hodgen, *Early Anthropology in the Sixteenth and Seventeenth Centuries*, Philadelphia (1964), esp. chs. 6—7; P. Harrison, '"Religion" and the Religions in British Thought: Lord Herbert to Hume' (Doctoral dissert., University of Queensland), Brisbane (1988), esp. chs. 4—5.
[30] Thus Westfall, '*Theologiae*', *loc.cit.*
[31] Keynes MS 3, pp. 5—7, 35.
[32] Trompf. *op. cit.*, pp. 301—2, cf. G. Bouwsma, *Concordia Mundi: the career and thought of Guillaume Postel (1510—1581)*, Cambridge, Mass. (1957), pp. 252—63, 283.
[33] For background, Lactantius, *Div. Inst.*, esp. I, x—xi; II, xiv; V, v, and for foreground, esp. J. Seznec, *La suivivance des dieux antiques*, London (1940). An important mediator of Lactantian Euhemeristic methods to Newton was G. T. Vossius, esp. in his *De Theologia Gentili, etc.*, Amsterdam (1641), Vol. 1 (which contains *De idolatriae origine ac progressu*). Newton read Vossius, certainly his *Rhetoricas contractae, etc.* Oxford, 1651, as an undergraduate; Westfall, *Never at Rest*, pp. 81—2, n. 46 on his notes.
[34] W. Raleigh, *The Historie of the World*, London (1614), Bk. 1, sig. F6ʳ—F6ᵛ (where the 'twelve severall gods' of the pagans are explained by Lactantian-style Euhemeristic references to Noah's family), cf. the polygenetic theories of others, e.g., following D. R. McKee on 'Isaac de la Peyrère; a precursor of eighteenth-century critical deists', in *Publications of the Modern Language Association*, **59**, 465 ff. (1944). For other positions in between (e.g., Vossius and G. Kirchmaier), D. C. Allen *The Legend of Noah*, Urbana (1963), pp. 88—9).
[35] Shuckford, *The Sacred and Prophane History of the World, etc.*, London (1727) (using 1743 edn., pp. 98—9).
[36] Defending Hebrew, note esp. the influential Postel, *De originibus seu Hebraicae linguae et gentis Antiquitate deque variarum linguarum*, Paris (1538), and among others in Newton's day the great bishop John Lightfoot, *Erubhin, or Miscellanies*, London (1629), after him. Defending Chinese (in the light of missionary research there), J. Webb, *An Historical Essay endeavouring a Probability that the Language of the Empire of China is the primitive language*, London (1669), cf. also Shuckford in his wake, *op. cit.*, p. 101.

[37] Jenkin, *The Reasonableness and Certainty of the Christian Religion*, London (1698) (using 1734 edn., vol. 1, pp. 47—8). Raleigh held the same views ahead of him, cf. *supra* n. 34 for reference.

[38] Yahuda MS 16.2, f.144 (Westfall reads 45).

[39] Sanhed. 56a—b, cf. also Yoma 28b. Newton would naturally have excluded the additional concerns while working on alchemical problems. Cf. A. R. and M. B. Hall, 'Newton's Chemical Experiments', in *Archives internationales d'histoire des sciences*, **11**, 113ff (1958), and cf. L. W. H. Hull, *History and Philosophy of Science*, London (1959), p. 119 on the traditional and crucial connection between alchemy and the making of optical glass, another key occupation of Newton's at this time.

[40] Maimonides, esp. *Mishn^e-Torah*, vol. 17 (*S^{e}fer Shoftim*, 89), London (1683) and earlier edns. (using Jerusalem (1972) Hebrew edn., pp. 404—5), cf. Manuel, *INH*, pp. 9, 42, etc.

[41] P. 382, Conveniently reproduced by Manuel in *INH*, plate 10, epp. p. 117, cf. pp. 239, 284—5. This scholium is curiously missed in Westfall's article on the 'Origines'. Since 'Benefactor' has a comma written after it, we can assume Newton projected a new English ending, which he put into his personal copy of the *Opticks*; cf. R. B. Webber (introd.), *A Descriptive Catalogue of the Grace K. Babson Collection of the Works of Isaac Newton, etc.*, New York (1950), p. 67. A general reference to the seven *Nondicharum praecepta* also appears at the end of Samuel Clarke's second latin edn. of the Opticks (1720) (cf. S. Horsley edn. of Newton's *Opera quae exstant omnia*, London (1782), Vol. 4, p. 264). Cf. also J. Selden, *De jure naturali*, London, 1640.

[42] In 'A Short Scheme of the True Religion', Newton lists in a chain Adam and Enoch before Noah, and then Abraham, Moses and Christ (McLachlin, *op. cit.*, p. 48). In Islam the sense of a chain of prophets was mediated through the *Isra'iyliyyat* (Israelite, largely Talmudic tales, which also had Samaritan influences to them). Cf. also *Qu'ran*, *Sur.* 2:119f.; 19:50ff. on a shorter listing of prophets preceding Muhammed.

[43] Quoted in Westfall, *Never at Rest*, p. 820. Note that bishop Edward Stillingfleet (1635—99) had written a widely read 'Oecumenically-Protestant' work by the title *Irenicum*, (London, 1661), perhaps reflecting the fact that, among all contemporary chronographers, Stillingfleet had a major influence on Newton; cf. ns. 103, 109, 115—6 *infra*.

[44] Using The Hague (1686) edn., pt. 1, pp. 442—8.

[45] The reference to Spencer in Yahuda MS 41, f. 5 dictates this placement; thus Westfall, *op. cit.*, p. 352, n.55; and for keeping pace with Spencer's researches, see *The Correspondence of Isaac Newton* (hereafter *Corres.*) (ed. H. W. Turnbull), Cambridge (1960), Vol. 3, p. 291 (to Mencke, 1693).

[46] 'Irenicum', in McLachlan, *op. cit.*, p. 28.

[47] For van Sleidan, using the 1624 Latin edn., *De quatuor summis imperiis*, Leiden, p. 1; for Lightfoot etc: in Newton, McLachlan, *op. cit.*, p. 127 (the *Horae*, with first volume [1658] on Matthew being published in Cambridge); for evidence of Newton's research into the Talmudists, e.g., Yahuda MS 2.4, f. 46, Babson MS 434, f. 1, cf. his 'Dissertation upon the sacred Cubit of the Jews and the Cubit of the Several Nations' (ca. 1680), later pub. in J. Greaves *Miscellaneous Works* (ed. T. Birch), London (1737), Vol. 2, pp. 421 (and n.), 425, 432. On Talmudic texts (and Maimonides) in the Cambridge libraries most accessible to Newton, esp. P. Gaskett, *Trinity College Library;*

the first fifty years, Cambridge (1980), p. 187, cf. P. Gaskett and R. Robson, *The Library of Trinity College, Cambridge, a short history*, Cambridge (1971), p. 10; S. Bush and C. J. Ramussen, *The Library of Emmanuel College, Cambridge, 1584—1657*, Cambridge (1986), pp. 56, 58—9, 61. Newton was learning Hebrew before 1661.

48 Spencer, *op. cit.*, pt. 1, p. 442. Newton's interest in specific portions of Spencer's work — note this partly dog-eared personal copy, J. Harrison, *The Library of Isaac Newton*, Cambridge (1978), p. 242, ct. 1545 — was for corroboration of his own *pre-existing* views, not for new ideas.

49 Conveniently printed in Manuel, *INH*, pp. 198—221 (App. 13). The tidiness of this MS, dated *ca.* 1693 (*Ibid.*, p. 198), suggests it was ready for publication.

50 *Ibid.*, p. 199.

51 See esp. Yahuda MS 16.1—2, fs. 23^v—24^v, cf. also *Corres.,* Vol. 3, p. 338 (1694) for another prospective title.

52 Philo, esp. *Vit. Mos*, ii, 12, etc., Josephus, *Adv Apion.*, ii, 154 [pseudo-]Justin, *Hort. adv. Graec.*, xii, xxvi, Clement, *Stromat.*, I, xvi, xxi, xxiv.

53 Josephus, *Antiq.*, I, viii, 2, using here William Whiston's translation (1737), from the London (1866) edn., p. 33; the relevant passage being referred to by Newton in the 'Origines', Yahuda MS 16. 12, f. 6^r.

54 Yahuda MS 16.1 f. 27^r (following 2, f. 23^v) (on Joseph, a reference to Marsham on the same folio); 'Original', p. 217 (quotation), cf. Manuel, *INH*, p. 120 (re arts, etc.).

55 Yahuda MS 16.2, fs. 3, 20ff., 25ff., 44. 48ff., cf. 16.1 fs. 15—6, 39, New College MS (at Bodleian Library, Oxford), II, f. 164 (pantheons), 'Original', pp. 206—7 (Cecrops), Yahuda MS. 16.1, f. 26 (or 16.2 f. 22^v); 16.2. f. 24, 'Original', p. 215 (Cadmus), cf. also Newton's (later) 'Remarks on the Observations made on a Chronological Index, etc.', in *The Philosophical Transactions of the Royal Society of London*, 7, p. 92 (1809), (whence short quotation), and for background, Clement, *Strom.*, i, xvi (an important source and passage for Newton).

56 Yahuda MS 16.2, f. 4 (using Clement, *Strom.* I, xvi), cf. also 'Original', pp. 218—9.

57 Yahuda MS, e.g., 16.1, fs. 3, 5, 7, etc. 16.2, f.22^v—24^r, etc. (and see n. 55 above), cf. Newton's (later) *The Chronology of Ancient Kingdoms Amended*, in *Opera*, Vol. 5, pp. 155—7.

58 New College MS II, f. 238 (cf. *supra*).

59 See esp. Yahuda MS 16.1, f. 39, 16.2 fs. 12, 16, 18. See ns. 32—3 *supra*, on Lactantius and Postel; for Kircher, see J. Goodwin, *Athanasius Kircher*, London (1979), p. 15; and cf. Vossius, *Theologia*, Lib. 1, pp. 118ff.

60 Yahuda MS 16.2, fs. 28—9, cf. fs. 16, 18, 42—3, 16.1, fs. 1—1^v, 7, 8, MS 17.2, fs. 2—2^v, and see Westfall, '*Theologiae*', pp. 20—1.

61 Yahuda MS 16.2, f. 1 (Saturn, etc.), 16.2, f. 12, cf. Lactantius, *Div. Inst.*, V, v (Janus), Yahuda MS 16.1, ff. 3, 9—9v (Jove, Belus, etc.) ff.

62 Eusebius (-Jerome), *Chronicorum* (ed. A. Schoene, Frankfurt (1875), Vol. 1), cols. 17ff., as cf. also Augustine, *De civit. Dei.*, esp. xviii, 2, and as background Justin, *Epit.* 1.2, Diodorus Siculus, *Bibliot.*, I, lvi, 5; II, i, 4—xx, 5.

63 Yahuda MS 16.1, fs. 6—7, 23, cf. 16.2, f. 20 (on Nimrod), there being some suggestions of this already in Petau, *Ration. temp.*, vol. 1, pp. 3—8. On the Nabataeans, Yahuda MS 16.1, fs. 3, 5, 7; and on scattered cities, esp. 'Original', pp. 199—200. On

Ussher making more of Nimrod as creator of a genuine empire, by contrast, see his *Annals, op. cit.*, p. 29 (Lat., p. 25).

[64] See Yahuda MS 16.2, fs. 7, 18, 53, 68; 17.2, f. 10, cf. Westfall, '*Theologiae*', p. 19 (an excellent summary). For Newton's sources, esp. Lactantius, *Div. Inst.*, II, xiv, cf. also Diodorus, *Bibliot.*, I, ix, 6; Lucian, *De Astrol.* (on which work see Yahuda MS 16.1, f. 12). See also Newton in New College MS II, f. 58 on his Euhemeristic interpretation of Ammon as the first astrologer.

[65] See Kircher, *Oedipus Aegyptica*, London (1652) [1655], *passim*; W.F. Perry, *The Children of the Sun*, London (1923); G. E. Smith, *The Diffusion of Culture*, London (1933) (cf. Trompf, *In Search of Origins*, London and New Delhi (1990), pp. 151—3).

[66] Yahuda MS 16.2, fs. 20, cf. 16.2, f.6. And also on Adonis in this connection, 16.2, f. 21.

[67] 'Original', p. 199.

[68] *Ibid.*, p. 203, yet cf. Polybius, *Hist.*, VI, vi, Iff.

[69] 'Original', p. 209.

[70] *Ibid.*, p. 220 (also with reference to Egyptian nomes).

[71] *Ibid.*, p. 201.

[72] *Ibid.*, cf. also *Observations upon the Prophecies of Holy Writ, etc.* (using *Opera*, Horsley edn., vol. 5), pp. 313—5. This Four Monarchy theory, derived in the long run from Jerome's *Comment. in Daniel.* (I, ii, 31—5; II, vii, 4—7), is Newton's 'sturdy historical frame' in comparison to Augustine's images of 'the two cities' and the two Babylons (thus against Manuel *INH*, p. 47, cf. Trompf, *Historical Recurrence, op. cit.*, pp. 222—5).

[73] 'The Original', esp. pp. 206, 220 (on prytenea), cf. p. 220 on priest kings, Yahuda MS 41, f. 8^r on priests, and on both again, New College MS III, f.65. See also Yahuda MS 17.3, fs. 7—8 on Vestal fire taken from the centre of the earth, and for an ancient source connecting Noah's sacrifice with a fire altar, Josephus, *Antiq.*, I, iv, 7. Vossius placed the first fire cult as early as Nimrod (*Theologia, op. cit.*, Vol. 1, p. 648).

[74] See ns. 32—3, 59 *supra* on sixteenth and seventeenth Euhemerism, cf. also Manuel, *The Eighteenth Century Confronts the Gods*, Cambridge, Mass. (1959), ch. 3. Key Euhemerists in Newton's Patristic sources were, of course, Clement, e.g., *Exhort.*, I; Lactantius (n.33), and Augustine, *De civ. dei.*, II, 5; IV, 27; VI, 18—27; XVIII, 8.

[75] Cf. e.g., Yahuda MS, esp. 16.1, fs. 28ff.; 16.2, fs. 25f.

[76] Cf., e.g., 'Original', p. 219, and on Egyptian astronomy, see *infra*, n. 137, and note David Gregory, in Newton's *Corres.*, Vol. 3, p. 384 (1694) on Egyptian 'Copernicanism'. Near Eastern prototypes of the Greek calendar were completely unknown to Newton (e.g., the tablet ^{mul}APIN^2, cf. B. L. van der Waerden, *Anfänge der Astronomie*, Groningen [1966], pp. 70ff.).

[77] E.g., M. Bernal, *Black Athena*, London (1987); P. Springborg, *The Chain of Benificence*, London (1989), chs. 2—6, cf. also. M. C. Astour, *Hellenosemitica*, Leiden (1967).

[78] Diodorus Siculus, *Bibliot.*, I, viii, 5—ix, 5; Lucretius, *De rerum natura*, III; Polybius, *Hist.*, VI, vi, 1—6; ix, 9; Eusebius, *Hist. eccles.*, I, ii, 18—20; iv, 4—6.

[79] As in *Letters relating to the Theory of Light and Colours* (in *Opera*, Vol. 4, p. 295), cf. *The Unpublished First Version of Isaac Newton's Cambridge Lectures on Optics*

1670—1672 (Camb. Un. Lib. MS. Add 4002), (introd. D. T. Whiteside), Cambridge (1973), p. 59.

[80] See Westfall, *Never at Rest*, esp. chs. 6—7. Barrow's work on optics has background interest here, cf. M. Roberts and E. R. Thomas, *Newton and the Origin of Colours*, London (1934), chs. 4—6.

[81] Cf. Newton's 'Of the [world to come], Day of Judgment and World to Come' (Yahuda MS 6, reprod. in Manuel, *INR*), pp. 128—9.

[82] *Philosophiae naturalis principia mathematica* (*Opera*, Vol. 2), esp. pp. 223—5, (Vol. 3), pp. 5—25, cf. also *Quaestiones quaedam philosophicae*, trans. in J. E. McGuire and M. Tamney (eds.), *Certain Philosophical Questions: Newton's Trinity Notebook*, Cambridge (1983), pp. 427—8.

[83] 'A short scheme, etc.', in McLachlan, *op. cit.*, p. 49 (quotation), cf., 'Four Letters to Richard Bentley', conveniently reprod. in M. K. Munitz (ed.), *Theories of the Universe*, New York and London (1957), pp. 211 ff., and see R. Thiel, *And There Was Light* (trans. R. and C. Winston), New York (1957), pp. 173ff (more generally).

[84] Cf. *Opticks*, pp. 217—20.

[85] Conduitt, Memorandum, King's College, Camb. (Keynes MS. 130. II0, cf. D. Kubrii, 'Newton and the Cyclical Cosmos: providence and the mechanical philosophy', in *Journal of the History of Ideas*, **28**, 340—5 (1967); and see also Manuel *INP.*, pp. 120, n. 20; 388. That Hooke had comparable cyclical ideas, see paper by Birkett and Oldroyd in the present volume.

[86] One may rightly suppose Newton's *Principia* to stand as a foil to Descartes' *Principia philosophicae* (1644), yet since Descartes' work includes both theological and general epistemological discussions (esp in pt. 1), there is food for thought in the more precise delimitations of Newton's title as the *mathematical* principles of a natural philosophy, leaving theology, and reflection on knowledge and truth in a general philosophical sense to Biblical and select (non-Trinitarian) Patristic ancient authorities. Origen, as astounding scholar and brilliant systematist, was Eusebius' hero (cf. *Hist. eccles.*, VI, xix; xxv; xxxii—xxxiv), Eusebius being one of Newton's favourite ancients. On Newton's most important appeals to Origen in connection with the Arian-Trinitarian tension, *Corres.*, Vol. 3, pp. 133—135, 138 (= Letter to a Friend, prob. John Locke, Nov. 1690), cf. also Westfall, *Never at Rest*; pp. 312—3, but a connecting of Origen with his own cosmology would have been wisely kept unpublicized (since Origen's theological works were condemned by orthodox theologians in the 390s). Cf. also *Corresp.*, Vol. 4, p. 403, n. 3. In contrast, on Newton's reaction to the Cartesian (mechanical) system of the world as productive of atheism, see *infra*, and n. 180.

[87] For details, esp. F. C. Haber, *The Age of the World: Moses to Darwin*. Westport, Conn. (1959), pp. 91—7.

[88] *Opticks.*, pp. 27—8.

[89] Exasperatingly Newton seems to avoid giving a response to Burnet's suggestions concerning the Flood in particular (as against Creation in general) in *Corres.*, Vol. 3, p. 326, cf. p. 334 (1680). On Burnet's views, *Telluris theoria sacra*, London (1680), pp. 190—6, cf. also P. Rossi, *The Dark Abyss of Time*, Chicago and London, 1984, ch. 7, and Gascoigne in this volume on Burnet's interests in Noah. On Whiston's views, *A New Theory of the Earth, from its Original to the Consummation of All Things, etc.*, London (1696), esp. pp. 305—7, 341—4, cf. Allen, *op. cit.*, pp. 107ff (also on Burnet).

[90] See R. W. V. Elliott, 'Isaac Newton's "Of an Universal Language"', in *The Modern Language Review* **52**, 1ff. (1957); and Newton's, 'Dissertation upon the Sacred Cubit', pp. 420, 425.
[91] See esp. Manuel, *INP*, pp. 110 ff. (cf. pp. 146—7 on Liège Jesuits opposing his theory of colours); *INR*, pp. 5—6, cf. pp. 67—8.
[92] Westfall, *Never at Rest*, pp. 481—7; C. Hill, 'Newton and his Society', in R. Palter (ed.), *The Annus Mirabilis of Sir Isaac Newton 1666—1966*, Cambridge, Mass., and London (1967), pp. 31ff.
[93] See J. Simon, 'The Two Angels', in *Transactions of the Leicestershire Archaeological and Historical Society*, **31**, 38—41 (1955). Note also W. Stukeley, *Memoirs of Sir Isaac Newton's Life* (1752) (ed. A. H. White), London (1936), esp. p. 38 on Clark as Henry More's pupil (and educated among Anglicans with still more Presbyterian sympathies), cf. Manuel *INP*, pp. 44—5, 51ff., 107—11. Manuel curiously misses the underlying argument of 'The Original' about conciliar government (in *INH*, ch. 8).
[94] 'Irenicum', in McLachlan, *op. cit.*, theses 2, 3, 5, 6, 11, 13 (pp. 38—41).
[95] Esp. Harrington's *A Discourse upon This Saying: the Spirit of the Nation is not yet to be trusted with Liberty, etc.*, London (1660), pp. 1—2, 10, 13—14.
[96] Conveniently reprod. in J.-F. Kahn, *History of Social Progress* (trans. J. White), London (1966), p. 57, pl. 47.
[97] *Per contra*, J.-B. Bossuet's, 1662 Sermon, as in *Oeuvres oratoires* (eds. C. Urbain and E. Levesque), Paris (1921), Vol. 4, p. 360—4; T. Hobbes, *Leviathan* (1651), using M. Oakeshott edn., New York and London, 1966, esp. chs. 30—1.
[98] Thus Ussher, *Annals, op. cit.*, p. [v], and p. 39 (Lat., p. 33).
[99] *A Short Chronicle* (in *Opera*, Vol. 5), p. 12; *Chronology*, p. 276. Ussher, *Annales* (Lat.), p. 32 does have Solomon's accession at 1015 B. C., cf. Petau, *Rat. temp., op. cit.*, p. 57 (different again).
[100] Manuel, *INH*, esp. ch. 6, cf. also Wilcox, *op. cit.*, pp. 208ff.
[101] Yahuda MS 16.1, f. 17, cf. Eusebius-Jerome, *Chron.* (Schoene ed.), Vol. 1, cols. 1ff.
[102] 'Original', p. 211.
[103] *Chronology*, pp. 28—46, cf. *Short Chron.*, pp. 3—7. Stillingfleet had comparable concerns; e.g., in *Origines sacrae*, London (1662), edn., pp. 15ff.
[104] 'Original', p. 221.
[105] Manuel, in *INH*, esp. pp. 98—122, has an excellent analysis. Cf. also n. 132 *infra* an corroboration from the 'Origines', which was unavailable to Manuel when he wrote *INH*. It was Manuel who first sensed that Newton had projected a great historical work (e.g., in *INH*, p. 16), but at a time, again, when the Yahuda MS was not available to reveal the true dimensions of the Newtonian enterprise.
[106] See *Corresp.*, Vol. 4, p. 316, n. 4, ch. Vol. 5, p. 354, n. 1.
[107] For some account of patristic usage, later reevaluations and modern textual reconstruction of Manetho's *Aegyptiaca*, W. G. Waddell, Introd., to Loeb ed., Cambridge, Mass. (1940), pp. xiv ff. *et passim*, cf. R. R. Bolgar, *The Classical Heritage and its Beneficiaries*, Cambridge (1958) on the Renaissance situation; and Rossi, *op. cit.*, chs. 22—3 on Manetho's importance in seventeenth century debates. For Manetho as 'counter-historian' to the Bible, A. Funkenstein, *Theology and the Scientific Imagination from the Middle Ages to the Seventeenth Century*, Princeton (1986), pp. 273—4.
[108] Because of I. Casaubon, *De rebus sacris et ecclesiasticis exercitationes, etc.*, London

(1614), esp. pp. 73—87. For Newton's own interest in the Hermetic tradition, B. J. T. Dobbs, 'Newton's *Commentary* on the *Emerald Tablet* of Hermes Trismegistis: its scientific and theological significance', in I. Merkel and A. G. Debus (eds.), *Hermeticism and the Renaissance*, Washington and London (1988), pp. 182ff.

[109] *Short Chron.*, p. 6 (on Manetho), cf., e.g., Stillingfleet, *Origines sacrae*, using Oxford (1797) edn., Vol. 1, pp. 39ff. The general argument is still run today by more conservative Biblical scholars, e.g., D. A. Courville, *The Exodus Problem and its Ramifications*, Loma Linda (1971), Vol. 1, pp. 166, 242f., 311, etc.

[110] *Chronology*, esp. pp. 31—50.

[111] *Short Chron.*, p. 12 (Ceres), 18 (Prometheus), 15 (Theseus), 21 (Dido).

[112] *Ibid.*, p. 14 (first quotation), 'Remarks on the Observations', *loc. cit.*, p. 92 (second two quotations).

[113] *Short Chron.*, pp. 14—16, cf. *Chronology* (*Short Chron.* attached), pp. 278—80. The crucial ancient source is Herodotus, *Hist.*, II, 102—11, and see Gascoigne in this volume for Marsham and Sesac.

[114] *Short Chron.*, pp. 15—16, cf. 14 (or 974 BC); *Chronology*, pp. 278—80. Cf. 2 Chron., 12:2—6. See also H. Kees, 'Sesostris', in *Paulys Realenkyk. class. Altert.*, Vol. 11 (2A, 2), pp. 1862—76.

[115] *Short Chron.*, p. 10, cf. New College MSS I, ff. 32—3. Note that Cadmus is the traditional founder of Thebes, and there linkable with Egypt (as well as Phoenicia). For background, Herodotus *Hist.*, II, 49, 145; Josephus, *Adv. Apion.*, I, 8; Clement, *Strom.*, I, xvi, cf. also Stillingfleet, *Origines*, pp. 20—1.

[116] For Josephus on Philostratus and other Phoenician recorders, *Adv. Apion.*, I, 144, etc., for Tatian on Theodotus, Hypiscrates and Mochus, *Ad. Graec.*, xxvii (cf. Newton, 'The Original', p. 214); for Eusebius on Philo of Byblos, *Praep. Evang.*, IV, ix, 22ff (cf. Newton, 'Origines', Yahuda MS 16.2, ff 9, 11, and also New College MSS III, f. 207^v), cf. also Theophilus of Antioch, *Ad Autol.*, xxii—xxiii; and Sanchoniathon, *apud* Eusebius, *Praep. Evang.* I and Eratosthenes, with R. Cumberland's translation, London (1720), being in Newton's library (Harrison, *op. cit.*, p. 231). Note also Stillingfleet, *op.cit.*, pp. 26—32 as giving comparable priority to Phoenician historians before Newton.

[117] *Short Chron.*, p. 10.

[118] *Ibid.*, p. 8, *Chronology* (*Short Chron.* att.), p. 272. For longer term background Josephus, *Adv. Apion.* I, 14, 75ff., etc. and shorter term, e.g., Ussher, *Annals*, pp. 7ff.

[119] Newton, 'The Original', p. 207; New College MS, II, f. 164, cf. A. W. Gomme, 'The Legend of Cadmus and the Logography', in *Journal of Hellenic Studies*, 33, 53ff., 223ff. (1913).

[120] See *Chronology*, pp. 63ff., Manuel, *INH*, pp. 73, 82 (cf. pp. 82—6 on the choice Newton had between Chiron and Pan), 87 (Clement), although Manuel does not note probable inspiration for Newton's interpretations in Vossius, *Theologia*, Vol. 1, p. 521.

[121] Esp. *Chronology*, p. 82, cf. Manuel, *INH*, pp. 73, 82 for MS sources of relevance.

[122] *Ibid.*, chs. 4—5 for the details. For Newton on absolute time in this connection, Wilcox, *op. cit.*, pp. 208—14.

[123] *Chronology*, pp. 74—5.

[124] *Short Chron.*, p. 19.

[125] *Ibid.*, p. 17, cf. New College MS II, f. 120^v (whence quotation).

[126] Thus H. Schliemann, *Troja: results of the latest researches, etc.*, London 1884, cf. *Oxf. Class. Dict.*, pp. 1097—8 for the modern dating of Troy's strata.

[127] Cf. R. Kohn, *False prophets*, Oxford (1986), pp. 36—9 (whence the phrase, which refers to Newton's results in physics, not his historiographical work).

[128] New College MS III, f. 189.

[129] *Prophecies*, pp. 297—302, cf. Hobbes, *Leviathan*, ch. 33. Note also I Sam. 16:1—13.

[130] Concurring with Manuel, *INR*, pp. 17, 21—2, 49, 63—4, 84ff, and the textual evidence cited there.

[131] Yahuda MSS 2.4, ff 17—40; 13.2, ff. 1—22; 14, ff. 1—8, 32—3; 28.5, ff. 1—3; Babson MS 434, ff.1, 41—3, 47—8, 54; cf. [the non-extant] *Lexicon propheticum* on the tabernacle (mentioned by Stukeley, *op. cit.*, p. 59), 'Dissertation', esp. p. 430; *Chronology*, pp. 236—44 (with plan); Westfall, *op. cit.*, p. 347 (earlier plan, in Babson collection). Of background importance to Newton's interests, esp. Henry More, as in *The Theological Works*, London (1708) edn. used, pp. 530—1; J. Lightfoot, *The Temple, especially as it stood in the days of our Saviour*, in *Works*, London (1684), Vol. 1 (with pp. 1064—7 on Solomon's temple, and p. 1049 with a plan). On the building of Solomon's temple in early seventeenth century mathematical writing, e.g. J. V. Andreae, *Collectanae mathematica decades xi*, Tubingen (1614), pl. 100.

[132] See Manuel, *INP*, pp. 164—5 on the possible association between the alchemical 'golden fleece' and the Argonauts. On Hiram in Newton, e.g. Yahuda, MS 16.1, f. 26 (or 2, f. 23ʳ); 'The Original'; pp. 213—4, *Chronology*, pp. 114, 155—7.

[133] *Ibid.*, pp. 175, 183. A joust at Manetho, among others.

[134] Josephus, *Adv. Apion.*, I, 13—14 (using Whiston's trans. p. 632).

[135] See Yahuda MSS 17.2, ff. 20—21; 41, f. 26, Keynes MS 130, cf. 'Original', p. 212 (and *Chronology*, p. 289).

[136] J. E. McGuire and P. M. Rattansi, 'Newton and the "Pipes of Pan"', in *Royal Society (Great Britain) Notes and Records*, **21**, 137 (whence quotation), cf. 127ff, (1966). Cf. also D. P. Walker, *The Ancient Theology*, Ithaca, NY. (1972).

[137] Josephus *Adv. Apion.*, II, 168, cf. I, 163 (on the notion of God and the Pythagoras/Israel connection) (Newton may well have concurred with Whiston, incidentally, that Josephus was a pre-Trinitarian (Ebionite) Christian; thus Whiston, in a dissert. following his trans., p. 671 [s.v. 1, iv—v]). On Pythagoras and Egypt, Newton, see n. 135 *supra*, yet cf. Josephus *Adv. Apion.*, I, 14 (both Egypt and Chaldaea), Thoth, *Corresp.*, Vol. 3, p. 338 (with background in Eusebius' *Praep. Evang.* I), and both Neptune and Plato, Yahuda MS 16.1 f. 15. On one listing by Newton of Greek philosophical and Biblical authorities on the omniscience of God, see Newton's marginal notes to the *General Scholium* as reprod. by McGuire and Rattansi, *loc. cit.*, p. 122, cf. *Mathematical Principles of Natural Philosophy* (ed. F. Cajori), Berkeley (1934), p. 545 (*General Scholium*).

[138] ULC (= University Library, Cambridge) Ad. MS 3965.6, f. 270ʳ; cf. also Newton's 'classical Scholia', Gregory MS (Royal Society, London,), f.8. Strabo placed Moschus prior to the Trojan War (*Geogr.*, XVI, ii, 24), others (not followed by Newton) identified him with Moses (e.g. Arcerius, edn. of Iamblichus, *De vit. Pythag.*, p. 33n., cf. also D. Sailor, 'Moses and Atomism', *Journal of the History of Ideas*, **25**, 3ff, (1964). For background to Newton's approach to atomism, McGuire and Rattansi *loc. cit.*, pp.

130ff.; McGuire, 'Atoms and the "analogy of nature"', in *Studies in History and Philosophy of Science*, 1970, 1ff. See also *infra*, and n. 175.

[139] Keynes MS 130, ULC MS Ad. 3970, f.619^r (dated *ca*. 1704), Gregory MS 247, f.13.

[140] *Chronology*, pp. 212—35, 245—63, cf. *Prophecies*, p. 312.

[141] See *supra*, and n. 72, cf. Trompf, *Recurrence*, vol. 1, pp. 222—3.

[142] For Newton's acknowledgements of Mede, esp. *Prophecies*, p. 464, 'Fragments from a Treatise in Revelation', in *INR*, p. 114.

[143] Thus *Short Chron.*, p. 27 (s.v. 412 [BC]: 'And here ends the sacred history of the Jews'). Narrative in the style of 1—2 Kgs. and 1—2 Chron is not picked up until 1 Macc (*ca*. 160 BC). Josephus, in *Antiq*., XI—XII, struggles to find materials to bridge the gap.

[144] For earlier commentaries on Daniel as an historical work, H. H. Rowley, *Darius the Mede and the Four world Empires*, Cardiff (1935), pp. 74ff.

[145] Thus Westfall, *op. cit.*, p. 81.

[146] ULC, Ad. MS 3987, f. 123. cf. the autographed MS of the *Chronology*, Ad MS 3988. See also Manuel *INH*, p. 163.

[147] *Prophecies*, pp. 480—1 (though p. 481 does mention the Muslim conquest of Constantinople in 1453).

[148] Cf., 'Fragments, etc', 'Of the World to Come', in Manuel, *INR*, pp. 107ff., 126ff.

[149] Manuel, *INR*, pp. 7, 99.

[150] 'Fragments', p. 119 (short quotation), 122 (long), cf. pp. 114—7.

[151] See ns. 4, 6 *supra*, cf. Hill, *loc. cit.*, esp. pp. 32, 37.

[152] Eusebius, *Hist. eccles* e.g., VII, xxv, 22—7 (quoting Dionysius); Origen, *De princip.*, e.g., III, iv; vi.

[153] *Prophecies*, p. 468, cf. p. 472. (on the two winds as the two Romes).

[154] *Corres.*, Vol. 3, pp. 129ff (to Locke? 1690). Note Rome's support of Athanasius, as cf. Athanasius. *Apol.*, ii, 20; iii, 50, etc.

[155] See n. 41 above, with related text, cf. R. de Villamil, *Newton the Man*, London, n. d. [1935], p. 18 (black pudding, animals); F. M. A de Voltaire quoted in Manuel, *INH*, p. vii (chastity).

[156] Yahuda MS 15, 3, f. 57, cf. Westfall, 'Theologiae', *loc. cit.*, p. 27.

[157] Esp. Keynes MS 3, pp. 5—7, and 'Irenicum' in McLachlan, p. 28.

[158] For the above, Yahuda MS 15.3, f. 57; 17.3, f. 12; 41, f. 5 (both quotations), cf. Westfall, *loc. cit.*, p. 27.

[159] 'Remarks on the Observations', p. 92.

[160] *Ibid*.

[161] Cf., e.g., Yahuda MS. 15.3, f. 57 (prophets), Yahuda MS 5.1, f. 7r; 25, fs. 20^r, 21^r (dating Christ's birth), *Corres.*, vol. 3, pp. 195, 214 (to Locke, on miracles), cf. Westfall, *Science and Religion in Seventeenth-Century England*, New Haven (1958), p. 207 on Newton's view of Jesus as the the last of the prophets (another slightly Muslim touch), cf. also [Anon.], 'A Dissertation on Sir Isaac Newton's Scheme for Setting the Chronology of our Lord's Ministry', in *The Family Expositor*, London (1739), Vol. 3, append. Note that Newton was evidently unaware of inter-Testamental statements about loving God and one's neighbour, as found, e.g., in Test. Isaac. 5:2, Test. Dan. 5:3, Test. Benj. 3:3, Jub. 20:9.

[162] Yahuda MS 7.2p. (no folio), cf. Acts 15.29, etc.

[163] For a pertinent quotation from the mathematician Wallis 'Three Sermons concerning the Sacred Trinity (1691), which in turn quotes Gales', Court of the Gentiles (Oxford, 1669—77, 2 Vols.), see McGuire and Rattansi, loc. cit., p. 133 (also on Cudworth).
[164] Cf. supra and ns. 17—20 on the social corruptibility of Egyptian Christianity. For ancient sources of a more general relevance to this, e.g., Script. Hist. Aug., VIII:1—8. Newton's persistence with Arianism into his later years is best reflected in his relations with William Whiston, his successor to the Lucasian Chair, cf. Westfall, Never at Rest, pp. 648—53, 700, 820 ff. On Whiston's more blatant anti-Trinitariansim and its consequences, e.g., Dict. Nat. Biog., vol. 21, pp. 12—3.
[165] Mede, The Key of the Revelation (here using London, 1656 edn.) pp. 114—25.
[166] For the evidence, Manuel, INR, pp. 19ff.
[167] See R. de Villamil, op. cit., p. 47 for the quotation.
[168] Keynes MS 7 (= with minor errors McLachlan, op. cit., p. 52).
[169] C. H. Josten (ed.), Elias Ashmole, 1617—1692, London, Vol. 1, pp. 33—5; D. C. Martin, 'Sir Robert Moray', in H. Harley (ed.), The Royal Society, London (1960), p. 246, cf. T. Sprat, History of the Royal Society, London (1667). Cf. also J. Hamill, The Craft: a history of English Freemasonry, London (1986), ch. 2.
[170] Following F. Yates, The Rosicrucian Enlightenment, London (1972), ch. 15. On Anderson as Talmudist, D. Knoop and G. P. Jones, A Short History of Freemasonry to 1730, Manchester (1940), p. 74.
[171] Store is set by the seven virtues (= the three Christian and four pagan virtues) being preserved orally. They are not in Anderson's Constitutions on Freemasons, London (1725), and the author has not found a Mason who will quote them all of a piece. See A. Horne, 'The Masonic Tradition and King Solomon's Temple', in Ars Quatuor Coronatorum, 80 8ff. (1967). See also F. Bacon, New Atlantis (1627), (in Essays, Civil and Moral, [ed. G. T. B.] Ward Lock edn. (1910) esp. pp. 482, 486 on the 'House of Solomon' (thought important as a paradigm for the founding of the Royal Society: D. McKie, 'Organ of the "New Philosophy"', in The Royal Society Tercentenary (from The Times supp., July 1960), London (1961), p. 22), R. Fludd, Utriusque cosmi historia, Oppenheim (1617—9), 2 Vols., cf. T. De Quncey, 'Historico-Critical Enquiry into the Origins of the Rosicrucians and the Freemasons' (1824), in his Collected Writings (ed. D. Masson), Edinburgh (1890), Vol. 13, p. 426 on the mystical architecture of Solomon's Temple (a theme as old as the Renaissance, cf. R. Wittkower, Architectual Principles in the Age of Humanism, London (1962), pp. 91, 106, 136, and see n. 131 supra). On connecting Noah, Solomon and Hiram, esp. The Graham MS, 1726, conveniently reprod. in Ars Quat. Coronat., 80, 70ff. (1967). On Hiram legends and Masonic initiation, J. la Fontaine, Initiation, Harmondsworth (1985), pp. 53, 57; W. O. Kaelber, 'Men's Initiation' in M. Eliade (ed.,) Encyclopedia of Religion, New York (1987), Vol. 7, pp. 232—3 (Hiram, Hiram legends and Masonic initiations). I owe to my colleague Harold Tarrant the suggestion that Newton was a critic of 'Masonic' insights, which probably represented a factional element in the Royal Society before his Presidency in 1703. Moves to prevent Whiston's membership of the Royal Society could have something to do with his possible association with Anderson and revised Freemasonry, cf. the former's revision of Anderson's A Genealogical History of the House of Yvery, London (1743), Vol. 2. For further background discussion, M. C. Jacob, The Radical Enlightenment, London and Boston (1981), pp. 116ff.

[172] On Newton's reserve and fear of criticism, esp. Manuel, *INH*, chs. 1—3.
[173] As we find in his exchanges with Robert Hooke, cf. esp. *Corresp.*, Vol. 3, pp. 40—44, 331—3.
[174] Thus W. C. Smith, *The Meaning and End of Religion* New York (1964), chs. 2—3; K. Thomas *Religion and the Decline of Magic*, Harmondsworth (1973), chs. 1, 6 *et passim*. For an early objectification of Catholicism and Protestantism as 'the same religion', cf. Archbishop Laud's appeal from prison, in *The History of the Troubles and Tryals of the Most Reverend Father in God and Blessed Martyr William Laud*, London (1695), p. 417, sect. 10.
[175] Justin Martyr, *Hort. ad Graec.*, xix (quotation). (Please note that this is now not usually considered to be a genuine statement by Pythagoras, although it occasionally appears in 'alternative' philosophical treatises of our century, e.g., M. P. Hall, *First Principles of Philosophy*, Los Angeles (1935), p. 41). See also Manuel, *INR*, pp. 23—4; McGuire and Rattansi, *loc. cit.*, p. 137, cf. Gregory, in Hiscock (ed.), *op. cit.*, p. 30 (s.v. 21 Dec 1705) for general background.
[176] For Newton on the inductive method, esp. *Principia* (in *Opera*, vol. 3) pp. 170—4. For discussion, R. Palter, 'Newton and the Inductive Method', in Palter (ed.)., *op. cit.*, pp. 244ff. On hypothetico-deductive elements in his 'natural-philosophic' method, however, see e.g., S. I. Vavilov, 'Newton and the Atomic Theory', in *The Royal Society, Newton Tercentenary Celebrations*, Cambridge (1947), p. 154; I. B. Cohen, *Franklin and Newton*, Philadelphia (1956), pp. 575—83; J. Kerival, *The Background to Newton's Principia*, Oxford (1965), p. 111; N. R. Hanson, 'Hypotheses Non Fingo', in *The Methodological Heritage of Newton*, Oxford (1970), pp. 14ff.; T. G. Cowling, *Isaac Newton and Astrology* (*Selig Brodetsky Memorial Lecture 18*), Leeds (1977), pp. 17—19.
[177] The phrase in Paul Ricoeur's; *The Reality of the Historical Past* (*Aquinas Lecture 1984*), Milwaukee (1984), p. 21, cf. Trompf, *Recurrence, op. cit.*, chs. 4—5; 'Macrohistory and Acculturation', in *Comparative Studies in Society and History*, 31/4 632f. (1989).
[178] Stukeley, *op. cit.*, p. 62; Fréret, *Défense de la chronologie, fondée sur les monuments de l'histoire ancienne, contre le système chronologique de M. Newton* (ed., J. P. de Bougainville), London (1758). Newton's world corresponded much more to the map crucial for ancient historians attached to L. E. Du Pin, *Bibliothèque unverselle des historians*, Amsterdam (1708) (a work he used for his *Chronology*) than the frontespiece map balancing the Americas and the Old World in Raleigh's *Historie*. Note his cooperation with B. Varenius (cf. *Geographic generalis . . . Ab Isaaco Newton, Math. Prof Lucasiano*, etc., Cambridge (1681), who did, after all, make a little of America (*De Diversis gentium religionibus*, in his *Descriptio Regni Japoniae et Siam*, Cambridge (1673), pp. 242—3).
[179] E.g., by Manuel, *INH*, pp. 137—8 193; Hill, *loc. cit.*, p. 41.
[180] Thus A. Koyré, *Newtonian Studies*, London (1965), p. 39, cf. also Rossi, *op. cit.*, p. 42.
[181] McGuire and Rattansi, *loc. cit.*, p. 138, cf. Manuel, *INR*, chs. 3, 5.
[182] Against Westfall, *loc. cit.* Cf. also McGuire and Rattansi, *loc. cit.*, pp. 120—1 on Newton's backward-looking extolling of the ancients (unlike Bacon). For general background here, e.g., R. F. Jones, *Ancients and Moderns*, St. Louis (1961).

NOTES ON CONTRIBUTORS

KIRSTEN BIRKETT is a research student in the School of Science and Technology Studies, University of New South Wales. She is currently working on science and religion in the seventeenth century.

JAMES FRANKLIN is senior tutor in mathematics at the University of New South Wales, and is currently writing a history of the early development of ideas about probability.

JOHN GASCOIGNE is senior lecturer in history at the University of New South Wales. He is the author of *Cambridge in the Age of the Enlightenment* (Cambridge: Cambridge University Press, 1989).

STEPHEN GAUKROGER is reader in philosophy at the University of Sydney. He is the author of *Explanatory Structures* (Sussex: Harvester Press, 1978), *Cartesian Logic* (Oxford: Clarendon Press, 1989), editor of *Descartes* (Sussex: Harvester Press, 1980), and has edited and translated *Arnauld: On True and False Ideas* (Manchester: Manchester University Press, 1990).

KEITH HUTCHISON lectures in the history and philosophy of science at the University of Melbourne. He is currently interested in social and political aspects of seventeenth-century science.

ALEXANDER JACOB is senior fellow at the Centre for Reformation and Renaissance Studies, University of Toronto. He is editor of *Henry More: The Immortality of the Soul* (Dordrecht: Kluwer, 1987).

JAMIE KASSLER is a visiting fellow in the School of Science and Technology Studies at the University of New South Wales. She is author of *The Science of Music in Britain, 1714—1830* (2 Vols., New York: Garland Press, 1979), and is one of the general editors of the works of Roger North. She is currently working on the role of music in understanding human cognition.

DAVID OLDROYD is associate professor in the School of Science and Technology Studies, University of New South Wales. He is author of *Darwinian Impacts* (Atlantic Highlands: Humanities Press, 1980), *The Arch Of Knowledge* (London: Methuen, 1986), *The Highlands Controversy* (Chicago: Chicago University Press, 1990), and he edited, with Ian Langham, *The Wider Domain of Evolutionary Thought* (Dordrecht: Reidel, 1983).

JOHN SUTTON is a research student in the Department of Traditional and Modern Philosophy, University of Sydney, working on the history of theories of the mind.

UDO THIEL is lecturer in philosophy at the University of Sydney. He is author of *Lockes Theorie der personalen Identität* (Bonn: Bouvier, 1983) and *John Locke* (Reihbeck: Rowohlt Verlag, forthcoming).

GARRY TROMPF is associate professor of religious studies at the University of Sydney. He is currently working on the second volume of his *The Idea of Historical Recurrence in Western Thought* (1st Vol. Berkeley: University of California Press, 1979).

INDEX OF

MYTHICAL AND HISTORICAL FIGURES

Abraham xi, 182, 185, 217, 233—4, 240n42
Abram 220, 222
Adam xi, 1, 193, 240n42
Adrian VI, Pope 138
Aeneas 6, 13—14
Agememnon 229
Alciato, Andreas 7, 22n36, 138
Alexander the Great 13
Alexander, Sir William, earl of Stirling 36
Alexander of Aphrodisias 31—2
Altdorfer, Albrecht 9, 20n22
Ames, William 97n17
Ammon 209
Anaxagoras x
Anderson, James 235
Angell, John 226
Aphrodite 222
Apollo 163, 1—9, 11, 15—6, 17n14, 18n9, 19n14, 23n39, 163
Apollodorus 6, 19n14, 22n32
Aquinas, Thomas St. 32, 82—3, 136—7, 139, 143n44—5
Argus 163—4
Aristotle x—xi, xvi n1, 5, 18n13, 26, 31—2, 46n49, 48n80, 53, 61, 53, 61, 65, 73n62, 77n106, 81, 114, 121n13, 133, 149, 154, 168n25, 169n54, 179
Arius 215
Arminius 42
Arnauld, Antoine 84, 86, 98n30, 125, 140
Artemis 17n4
Ashmole, Elias 235
Athanasius 215—16, 232, 234, 247n154
Atlas 14—15, 164—5

Augustine St. xiii, xv, 26, 22n34, 27, 30, 44n10, 82, 96n10, 222, 241n62, 242n74
Aurelias, Marcus 62, 74n68
Azo, the Glossator 135

Bacchus 228
Bacon, Francis x, 37, 49n102, 145, 155, 169n61—3, 17, 248n171, 249n249
Baldus de Ubaldis 128—9, 145n15
Baro, Peter 42
Barrow, Isaac 214—15, 238n21, 243n80
Bartholome de Medina 137
Bathurst, Ralph 71n10
Bauny, P. 130, 142n23
Bayle, Pierre 30, 46n34, 179, 207n32
Begar, Laurentius 9, 20n22
Belvus (Assyrian King) 222
Berkeley, Abraham 167n16
Berkeley, Bishop George 101, 107, 115—19, 121n22
Bernoulli, James 124, 141
Boccaccio, Giovanni 23n37
Bocchi 23n37, 23n41
Bo[u]ssuet, Bishop J. 197, 211n96, 244n97, 211n96
Boyle, Robert 171
Brahe, Tycho 177
Bramhall, John 33, 36—9, 43, 46n59, 48n87—8, 49n95, 54—5, 70n5
Browne, Peter 80
Bruno, Giordino 180, 195
Burman, Frans 85
Burnet, Thomas 153—4, 169n47, 180—6, 189—90, 195, 198, 208, 208n9, 208n33, 208n42, 208n45, 209n54, 225, 243n89
Burthogge, Richard 99n48, 106—7, 120n11

253

Cacus 6, 14, 17n4
Cade, Anthony 84, 97n17
Cadmus of Tyre 221, 228—30
Calvin, Jean 26, 29, 42, 45n28, 47n65, 50n134
Cardano, Nicholas 39, 125
Cartari, Vicenzo 7, 19n20, 20n21, 21n25
Casaubon, Isaac xii, 244n108
Cecrops 221, 223
Ceres 165, 228, 245n111
Charles I 42
Charles II 213
Charles V, Emperor 129
Chevalier de Mere 124
Chillingworth, William 139, 140
Chiron 229, 245n120
Christian, Thomasius 86
Chrysippus 27, 44n8, 44n11, 44n13, 55
Chrysostom, Dio 18n12, 22n32
Cicero 12—13, 21n30, 44n12, 73n63, 77n106, 133, 234
Clarke, Samuel 80, 96n6, 240n41
Clauberg, Johannes 86
Clement of Alexandria 173, 206n4, 229, 242n74
Coke, Edward 138, 144n52
Collins, Anthony 80
Conduitt, John 219, 225, 232, 242n85
Confucius 186, 234
Constantine 232
Copernicus, Nicholas ix, 1—3, 7, 9, 15—16, 18n10, 19n19, 23n38, 41, 156, 169n68
Cordemoy, Gérard de 86
Cowley, Abraham x, 55, 65—7, 71n10
Cromwell, Oliver 226
Cudworth, Damaris 95
Cudworth, Rafe 43
Cudworth, Ralph xii—xiii, xvin2, xvin4, 27, 30, 33—4, 39, 41, 43, 47n73, 50n136, 51n138—9, 51n146, 79—82, 86—95, 95n1, 98n25, 98n38—9, 101, 107, 111—15, 121n14, 192, 209n70, 210n78, 210n79, 216, 234, 248n163
Cumberland, Richard 80, 96n4

Cush (Noah's son) 221

Daphne 163
David (biblical) 2, 17n4, 194, 227—8, 230
Democritus xi
Descartes, René 56, 80, 84—6, 89—90, 101, 243n86, 50n122, 97n19—21, 98n25, 98n38
Deucalian 163
Dido 228, 245n111
Digby, Sir Kenelm 28, 45n20
Digges, Thomas 23n41
Diodorus of Sicily 14, 19n14, 19n17, 22n30, 22n33—4, 155, 164, 224, 241n62, 242n78
Dionysos 13, 19n18
Dodwell, Henry 96n5, 214, 238n9
Donne, John 41—2, 50n128
Dryden, John 237n2
Du Choul, Guillaume 10, 20n23, 22n33
Du Moulin, Charles 129, 142n18

Edwards, John 178—9, 181, 207n20—1, 207n23, 208n30
Eli 194
Elizabeth I, Queen 139
Empedocles x, 113
Erasmus, Desiderius 30, 46n38
Eratosthenes 149, 167n20, 245n116
Euhemeris of Messene 155
Eumenius 7—8, 18n11
Eumolpus 6
Eusibius of Caesarea 214, 222, 224, 227, 232, 241n62, 243n86, 245n116, 246n137, 247n152
Evander, King 6
Eve 183, 193
Evelyn, John x

Fabianus, Servius Flavius Papirius 162, 170n74
Fermat, Pierre de 124—5, 131, 140, 143n25
Ficino, Marsilio xiii, 39, 173, 120n12
Forge, Louis de la 86
Francis, Alban 226

INDEX 255

Fraunce, Abraham 15, 22n31, 22n37

Gaforus, Franchinus 2, 7
Gale, Theophilus 174, 191, 206n7, 234
Galileo Galilei 74n69, 125, 133, 171
Gassendi, Pierre 179
George, St. 2
Geryon 15
Gesner, Conrad 167n16
Geulincx, Arnold 86
Gibbon, Edward 199
Giovio, Paolo 32, 46n55
Giraldi, Lilio 7, 19n19, 20n20, 22n37
Glanville, Joseph x—xiii, xvin1
Goclenius, Rudolph 83, 97n13, 97n16
God xv, 15, 27, 29—33, 38—42, 47n65, 50n122, 53—4, 64, 66, 69, 76n102, 87—9, 104, 107, 111—12, 118, 120n12, 165—6, 171—2, 174—9, 181—3, 185—8, 190, 201—2, 204—5, 215, 217—18, 223, 230
Golding, Arthur 152, 168n26, 168n31, 168n35
Gomarus 42
Gorgias x
Gratian 135
Gregory, David 191—2
Greville, Fulke 28, 45n21—2
Grotius, Hugo 138, 144n56

Hardian, Emperor 136
Hale, Sir Matthew 87, 98n33, 153, 169n51
Halley, Edmond 146—7
Ham 190, 209n70, 221
Hammond, Henry 97n17
Hanno the Carthaginian 149—50, 154, 167n16—18
Harrington, James 226, 244n95
Harris, John 80
Harvey, William 55, 59—62, 71n13, 72n39, 72n42—3, 72n47—8, 73n59—60, 73n63, 75n78
Hebe 15
Heber xi
Helios 11
Henry VIII, King 139

Heraclitus x, 19n14, 113, 116
Herakles 2—16, 17n3, 18n8—10, 19n18—9, 20n22, 21n27, 23n39
Herbert of Cherbury 84, 97n18
Hercules 10, 15, 228
Hermes xiii, 6—7
Hermes Trismegistus 173
Hero 15
Herodotus 154, 169n53, 228, 245n113—5
Hesiod 150
Hesperus 15
Hevelius 166n6
Hiram of Troy, King 230, 246n132
Hobbes, Thomas 26, 32—9, 43, 47n78, 48n85—6, 48n89, 48n92, 53—9, 61—9, 70n1, 70n3, 70n5, 70n7, 71n12—13, 71n16, 71n23, 71n26, 72n39, 72n44, 73n61, 74n67, 74n69, 74n71, 74n74—5, 75n76, 75n78, 76n89—90, 76n97, 76n99—100, 76n102, 77n103—4, 77n106, 77n115, 78n117—18, 85—6, 90, 96n10, 99n40, 101, 216, 226, 244n97, 246n129
Homer xiii, 22n32
Hooke, Robert 145—54, 156—62, 165, 166n3—6, 167n10—11, 167n16, 167n18, 167n20, 168n25—6, 168n31, 169n52, 169n66, 170n71—2, 243n85, 249n173
Hooker, Richard 41, 50n129
Horace 22n32, 23n41
Horn, Georg xi
Hudde, Johann 125, 130
Huygens, Christiaan 124—5, 130—2, 140
Hylobares 121n18

Innocent III, Pope 135—7
Io 163
Isis 23n40, 228

James II, King 226
James, Thomas 28
Janus 222, 241n61
Jenkin, Robert 217, 240n37

INDEX

Jerome, St. 222, 227
Jesus xiv, xv, 186, 191, 198, 200, 204, 215, 217, 228, 233—4, 240n42
Job ix
Joseph 174, 221
Josephus 220, 228, 230, 241n52—3, 242n73, 245n116, 246n137, 247n143
Jove 222, 241n61
Juno 15, 163—4
Jupiter 15, 162—3, 222
Justinian, Emperor 126

Kepler, Johannes 133, 171, 177
Kircher, Athanasius 150, 153, 168n28, 169n50, 221—2, 241n59, 242n65

Lactantius xiii, 222, 239n33, 241n59, 242n74
Laney, Benjamin 36
Laplace, Marquis de 26
Laud, William 41
Le Clerc, Jean 95, 190
Leibniz, Gottfried Wilhelm 80, 95, 99n47, 124—5, 133, 138, 140—1, 143n64, 143n67
Lessius 130, 132, 138, 142n22
Liber 13
Lightfoot, John 219, 239n36, 246n131
Linos 5—6
Lippold, Georg 17n4
Lipsius, Justus 26—8, 34, 44n8, 44n10—11, 45n25, 70n6
Lister, Martin 146
Livy 13, 22n32, 142n6
Llwyd, Edward 146
Locke, John 95, 161, 178, 247n154, 247n161
Lucan 12—13, 21n30
Lucian 7
Lucretius 41, 224
Lugo, Juan de 130—1, 142n24
Luther, Martin 26, 30—1, 46n40, 47n65
Lycaon 162—3
Lycurgus 220

Macrobius 5, 10—11, 18n11, 20n24, 21n25
Maffei, D 142n19
Maimonides 218, 240n40
Malebranche, Nicholas 86
Manetho 227—8
Manilius, Marcus 12—13, 21n30
Marcion xiv
Mars 228
Marsham, John 197—200, 203, 211n99—100, 214, 221, 238n9, 245n113
Marston, John 25, 42, 45n23—4, 49n98—9
Martyr, Justin 206n4, 220, 235, 249n175
Marsyas 6
Mascardi, Giovanni 138, 144n51
Maxwell, John 80—1
Mayne, Zachary 96n3
Mede, Joseph 213, 231—2, 234, 237n6, 247n142
Medusa 165
Melanchthon, Philipp 30
Melissus x
Mencke, Otto 190
Menes 230
Menochio, Giovanni 137, 143n49
Mercury 7—8, 15—16, 19n17, 163
Mersenne, Marin 27, 34, 39—41, 48n79, 50n116, 50n122, 56
Micraelius, Johannes 83
Middleton, Conyers 199, 202—3, 211n106—7
Milton, John 25, 44n3, 48n93
Monboddo, Lord 204
More, Henry 48n87, 79—80, 88, 101, 103—12, 115, 117—19, 119n1, 120n4—5, 120n7—9, 120n12, 121n13, 192, 244n93, 246n131
Moray, Robert 235
Moschus the Phoenician 192, 231
Moses 175, 182—6, 191—3, 196, 198, 200, 217, 220, 233—4, 246n138

Nashe, Thomas 43, 137n51

INDEX 257

Neptune 17n4, 231, 246n137
Nero 161
Newton, Isaac ix, 116—17, 120n12, 121n23, 124, 171, 173, 184—98, 201—4, 209n48, 209n51, 209n57, 209n59, 210n76, 209n78, 211n89, 211n107, 213—26, 228—36, 237n5, 238n13, 239n33, 239n36, 240n39, 240n41—3, 240n47, 241n48, 241n53, 241n55, 241n57, 242n64, 242n72, 242n74, 242n76, 243n86, 243n89, 244n90, 244n105, 205n109, 245n116, 245n119—20, 245n122, 246n127, 246n132, 246n137—8, 247n142, 247n161, 248n164, 248n162, 249n176, 249n178, 249n182
Nicole, Pierre 140
Nimrod 222, 241n63, 242n63, 242n73
Noah 152, 182, 185, 187—90, 192, 197, 217—18, 220—2, 225—6, 233, 239n34, 240n42, 242n73
Nobilior, Fulvius 4, 8
Nonnos 11, 20n24, 22n32, 23n40

Oldmixon, John 184
Origen of Alexandria 225, 232, 243n86, 247n153
Osiris 13, 23n40, 228
Ovid 4—5, 12—13, 18n10—11, 21n28, 147—8, 150—2, 154—6, 162, 166, 168n31, 168n34, 168n36, 169n75

Palaephatus 155
Pan 7
Pascal, Blaise 124—5, 130—2, 140, 143n25
Patricius, Franciscus xvin1
Pegasus 165
Pemberton, Henry 237n2
Perkins, William 42, 97n17
Perry, John 222
Perseus 2, 17n4, 164—5
Petau, Denis 214
Phaer, Thomas 13
Phaeton 12, 164

Philo Judaeus 220, 241n32
Philolaus 156
Philostratus 22n36—7
Philotheus 121n18
Phoebus 12, 164
Pico della Mirandola 25, 39, 43
Pignoria Lorenzo 8, 19n20
Pilkington, John 50n133
Plato x—xv, 5, 18n13, 81, 114, 147—50, 154—5, 166, 189, 220, 246n137
Plautus 38
Pliny the Elder xi, 147—9, 160, 168n21—2
Plot, Robert 146
Plotinus xii—xiii, 80, 87—9, 93—4, 98n37, 99n45, 101—3, 109, 114, 120n2—3
Plutarch 5—6, 11, 13, 18n13, 19n14, 18, 21n26, 22n32—3, 23n40
Pluto 231
Polybius 224
Pompanazzi, Pietro 25—6, 30—5, 39—40, 44n4, 46n46, 46n48, 46n50, 47n63—4, 47n67—71, 48n81, 48n85, 49n111
Pomponius Mela 149
Porphyry xiii
Postel, Guillaume 217, 221, 239n36, 241n59
Priam of Troy 229
Prometheus 228, 245n111
Proserpine 165
Protagoras x
Ptolemy 3, 170n72, 193, 214
Pyrrha 163
Pythagoras xi, 154, 230—1, 235—6, 246n137, 249n175

Rale[i]gh, Sir Walter 155, 169n60, 216, 239n34, 240n37
Reynolds, Henry 155, 169n64
Robertson, William 204—5, 212n119
Romulus 223
Ross, Alexander 14, 20n20, 22n36

Salutati, Collucio 22n36, 23n37

258 INDEX

Samian 235
Samuel 194, 230
Sanchoniathon 245n116
Sanderson, Robert 97n17
Sanehedrin 218–9
Santerna, P. 142n19
Saturn 162, 222, 241n61
Scaliger, Joseph 214, 238n13
Sem xi
Semiramus 222, 231
Seneca 29, 45n24, 148, 161–2, 170n74
Sergeant, John 99n48
Servetus, Michael xv
Servius 11, 14, 22n34
Sesac 173, 198, 228, 245n113
Sesostris 173, 198, 228–9
Seutonius 5, 18n11, 142n6
Sherlock, William 99n48
Shishak 228
Shuckford, Samuel 217, 239n35–6
Sidney, Sir Philip 38
Smith, Grafton Elliott 222
Smith, John 95n1, 99n41
Socrates 186, 234
Solomon 173, 194, 211n107, 226–8, 230, 244n99, 246n131
Soto, Domingo da 129, 142n20
South, Robert 99n48
Spencer, John 175–80, 184, 187, 189–91, 195, 198–9, 201, 203, 206n9, 207n10–18, 211n103, 219, 240n45, 241n48
Spinoza, Baruch 98n28
Sprat, Thomas x, 155–6, 169n65, 248n169
Steno, Nicolaus 166n1
Stephen of Byzantium 167n16
Stillingfleet, Bishop Edward 197, 211n98, 214, 238n9, 240n43, 244n103, 245n109, 245n115–6
Strabo 148–9, 161, 167n20, 168n20, 168n24, 169n58
Straccha, B. 142n20
Stukeley, William 236, 249n178
Suárez, Francisco 138

Tacitus 29

Tatian 228
Taylor, Jeremy 97n17
Theophilus of Antioch 228
Theron, King 11
Theseus 228, 245n111
Theseus 2
Thoth 231
Tiberius Ceasar 161
Tillotson, John 184
Tindal, Matthew 198–99, 203, 211n102
Toland, John 198, 200, 211n101
Tribonian 126
Tromp, Cornelius 213
Tschirnhaus, Ehrenfried Walter von 86
Turner, John 99n48
Turnus 13
Twyne, Thomas 13

Ulpian 126–7
Urania 6
Ussher, James 214, 226, 238n13, 242n63

Valla, Lorenzo 30–1
van Leeuwenhoek, Anton 73n60
van Schooten, Frans 131
van Sleidan, Johannes 213, 219
Vanini, Giulio Cesare 33–4, 39
Varro, Marcus Terentius 168n33
Venus 222
Vico, Enea 9, 20n22
Virgil 6, 12–15, 21n28–9, 22n37, 148
Vitoria Francisco de 138
Voltaire, François-Marie Arouet de xi, 197, 203, 211n97, 247n155
Vossius, Gerardus 7, 19n19, 22n36–7, 221, 239n33–4, 241n59, 242n73, 245n120

Waller, Richard 147, 166n7
Wallis, John 146, 147, 166n116, 174, 206n3, 234, 247n163
Warburton, William 200–3, 211n108, 212n110–4
Warren, Erasmus 153, 169n48
Waterland, Daniel 199, 203, 206n4

Webb, J. 239n36
Webster, John 25, 29, 37—8, 41—2, 44n1—2, 45n26, 49n104, 49n111, 50n127
Whiston, William 153, 169n99, 189—90, 209n64, 225, 241n53, 243n89, 246n137, 248n164, 248n171
Wilkins, John 87, 96n10, 98n34
Witsius, Hermann 178
Witt, Jan de 125, 130, 132, 140

Wolff, Christian 80, 95n2
Woodward, John 153—4, 169n52, 180—2, 207n24—6

Xenocrates x
Xenophon 18n12

Zaltieri, Bolognino 20n20
Zeno 113
Zoroaster xi

AUSTRALASIAN STUDIES
IN HISTORY AND PHILOSOPHY OF SCIENCE

General Editor:
R. W. Home, *University of Melbourne*

Publications:
1. R. McLaughlin (ed.): *What? Where? When? Why?* Essays on Induction, Space and Time, Explanation. Inspired by the Work of Wesley C. Salmon. 1982 ISBN 90-277-1337-5
2. D. Oldroyd and I. Langham (eds.): *The Wider Domain of Evolutionary Thought.* 1983 ISBN 90-277-1477-0
3. R. W. Home (ed.): *Science under Scrutinity.* The Place of History and Philosophy of Science. 1983 ISBN 90-277-1602-1
4. J. A. Schuster and R. R. Yeo (eds.): *The Politics and Rhetoric of Scientific Method.* Historical Studies. 1986 ISBN 90-277-2152-1
5. J. Forge (ed.): *Measurement, Realism and Objectivity.* Essays on Measurement in the Social and Physical Science. 1987
ISBN 90-277-2542-X
6. R. Nola (ed.): *Relativism and Realism in Science.* 1988
ISBN 90-277-2647-7
7. P. Slezak and W. R. Albury (eds.): *Computers, Brains and Minds.* Essays in Cognitive Science. 1989 ISBN 90-277-2759-7
8. H. E. Le Grand (ed.): *Experimental Inquiries.* Historical, Philosophical and Social Studies of Experimentation in Science. 1990
ISBN 0-7923-0790-9
9. R. W. Home and S. G. Kohlstedt (eds.): *International Science and National Scientific Identity.* Australia between Britain and America. 1991 ISBN 0-7923-0938-3
10. S. Gaukroger (ed.): *The Uses of Antiquity.* The Scientific Revolution and the Classical Tradition. 1991 ISBN 0-7923-1130-2

KLUWER ACADEMIC PUBLISHERS – DORDRECHT / BOSTON / LONDON

Printed in the United States
17469LVS00002B/77